Collins · *do brilliantly !*

Instant**Facts**

Physics

A-Z of **essential facts** and definitions

Eric Deeson

William Collins' dream of knowledge for all began with the publication of his first book in 1819. A self-educated mill worker, he not only enriched millions of lives, but also founded a flourishing publishing house. Today, staying true to this spirit, Collins books are packed with inspiration, innovation and practical expertise. They place you at the centre of a world of possibility and give you exactly what you need to explore it.

Collins. Do more.

Published by Collins
An imprint of HarperCollins*Publishers*
77–85 Fulham Palace Road
Hammersmith
London
W6 8JB

Browse the complete Collins catalogue at

www.collinseducation.com
© HarperCollins*Publishers* Limited 2005

First published as Collins Gem Basic Facts Physics 1982

10 9 8 7 6 5 4 3 2 1

ISBN 0 00 720513 9

British Library Cataloguing in Publication Data
A catalogue record for this publication is available from the British Library

Every effort has been made to contact the holders if copyright material, but if any have been inadvertently overlooked, the Publishers will be pleased to make the necessary arrangements at the first opportunity.

Edited and Project Managed by Marie Insall
Production by Katie Butler
Design by Sally Boothroyd/Jerry Fowler
Printed and bound by Printing Express, Hong Kong

You might also like to visit
www.harpercollins.co.uk
The book lover's website

Introduction

Instant Facts Physics is one of a series of illustrated A–Z subject reference guides of the key terms and concepts used in the most important school subjects. With its alphabetical arrangement, the book is designed for quick reference to explain the meaning of words used in the subject and so is an excellent companion both to course work and during revision.

Bold words in an entry identify key terms which are explained in greater detail in entries of their own; important terms that do not have separate entries are shown in *italic* and are explained in the entry in which they occur.

Other titles in the *Instant Facts* series include:
English
Modern World History
Biology
Science
Geography
Maths
Chemistry

A

absolute zero The lowest temperature possible, when all thermal energy has been removed. It is zero on the **Kelvin scale**: 0 K = –273 °C. (Note that on the Kelvin scale no degree sign is used.)

absorption 1. The taking up of one substance by another; for example, a sponge takes up (*absorbs*) water.
2. The taking in of radiation by a substance; for example, a red filter absorbs all colours of light except red; gamma rays are absorbed by lead; and infrared radiation is absorbed by water vapour in the atmosphere, or by dark clothing. The effect of this kind of absorption is usually to heat the absorber.
3. The taking in of **fundamental particles** by a substance; for example, neutrons are absorbed by the material of the control rods in a nuclear reactor.

absorption spectrum *See* **spectrum**.

acceleration (a) An object's change of **velocity** *v* in unit time. The SI unit (*see Appendix A*) is the metre per second per second, m/s². The acceleration that takes place during a time interval *t* is:

$$a = \frac{v_2 - v_1}{t}$$

where v_2 is the final velocity and v_1 is the initial velocity.

Because velocity involves both speed and direction of movement, an object accelerates if either of these changes. *See also* **centripetal force**.

An object will accelerate only if a **force** is applied. For an object of mass *m*, the acceleration produced by a force *F* is:

$$a = \frac{F}{m}$$

The magnitude of the acceleration of an object is the slope of its speed/time graph at any moment. Negative acceleration is sometimes called *deceleration* or *retardation*.
See also **equations of motion, free fall, Newton's laws of motion**.

accelerator A machine that uses electrical forces to accelerate charged particles of matter (for example, electrons, protons and ions) to a very high speed. In large modern machines the particles are guided by magnetic fields. Accelerators are used for research on the **fundamental particles** of matter. The largest are in Switzerland (operated by CERN, the European Laboratory for Particle Physics), the USA and Russia. These machines take the form of *colliders*, in which beams of particles are stored for many hours and then caused to collide head-on, to gain higher interaction energies. *See also* **particle physics**.

accuracy An indication of how close a measurement is to the correct value. It is important to quote numerical answers to the number of figures that are likely to be correct, taking into account the expected accuracy of any information used or experimental readings taken. Thus in an experiment to find **specific thermal capacity**, temperature readings might be 24.0°C and 48.5°C (reading to the nearest 0.5°C). However, experimental **errors** might cause the readings to be correct to the nearest degree only. In the same experiment, mass readings could be 79.24g and 210.42g. The accuracy of the result is limited to the accuracy of the poorest readings – in this case the temperature. The results of the whole experiment can be quoted to only two **significant figures** at best, especially as errors accumulate.

After processing readings with a calculator, the 7- or 8-figure answers obtained must be reduced to two significant figures (in this example) and should preferably be given in **standard form**.

Accuracy can be improved in an experiment by:
(a) repeating readings many times and taking the average (mean);
(b) taking steps to reduce expected experimental errors, modifying the experiment, and so on;
(c) using more sensitive equipment – a micrometer gauge instead of a ruler, for instance.

acid A substance able to provide hydrogen **ions** (H^+ ions) for chemical reactions. An important type of reaction is that between an acid and a **base**: the hydrogen ion (which is the same as a **proton**) transfers from the acid to the base. The typical reaction is:

$$acid + base \rightarrow salt + water$$

$$HNO_3 + NaOH \rightarrow NaNO_3 + H_2O$$

acoustics **1.** The science of **sound**.
2. The characteristics of a building or space as it responds to sound – for example, whether some frequencies are more strongly reflected or absorbed than others, and how long it takes a sound to die away as a result of repeated echoes.

action An outdated word for **force**.

activity The number of particle emissions (alpha, beta, gamma) per second from a radioactive source. Activity is measured in *becquerels* (Bq), named after the physicist who discovered **radioactivity**, Henri Becquerel.

adhesion Attraction between particles of different substances. For instance, adhesion between water and glass particles causes the **meniscus** shown in the diagram.
  See also **capillarity, cohesion, kinetic model, surface tension.**

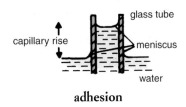

adhesion

air pressure *See* **atmospheric pressure**.

alkali *See* **base**.

alpha decay The disintegration of certain radioactive **isotopes** with the release of an **alpha (α) particle**. The **proton number** decreases by two and the **nucleon number** by four as a consequence. For example, radium (proton number 88, nucleon number 226) decays into radon gas (proton number 86, nucleon number 222):

$$^{226}_{88}\text{Ra} \rightarrow {}^{222}_{86}\text{Rn} + {}^{4}_{2}\text{He}$$

The $^{4}_{2}$He nucleus is the α particle. Energy is released, most of it being carried away as the kinetic energy of the alpha particle, which moves off at high speed. *See also* **atomic radiations, beta decay**.

alpha (α) particle The most massive of the particles emitted by the **radioactive** decay of an atomic nucleus. An alpha particle is identical to the nucleus of a helium atom, consisting of two neutrons and two protons. It leaves the parent nucleus at high speeds, having gained kinetic energy from a loss in mass of the nucleus. Alpha particles interact easily with matter, because of the double electric charge they carry. This means that they rapidly lose kinetic energy by colliding with and ionizing atoms or molecules in any material they travel through. This in turn means that they have a very short range in air (a few centimetres) and are stopped completely by even a thin piece of paper. *See also* **atomic radiations, radioactivity**.

alternating current (ac) A continuous electric current that regularly reverses its direction. Seen in wave form, its value varies regularly around zero. The mains alternating current in the UK is a sine wave, as shown in the diagram. It has a frequency of 50 Hz (hertz, or cycles per second).

Note that the term 'ac' is often used loosely – as in 'ac voltage', for instance, rather than 'alternating voltage'. *See also* **rectifier**, **generator**.

alternative energy The name loosely used for energy obtained from sources that are renewable and (in theory) non-polluting. Examples of alternative energy sources are wind generators, solar panels, geothermal energy (hot rocks), wave machines, and quick-growing wood.

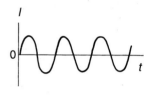

alternating current
Graph of current (I) against time (t).

alternator *See* **generator**.

ammeter A device used to measure electric **current**. The most common types are the *moving-coil* and *moving-iron* **meters**, though electronic meters (digital meters) are becoming more widely used. In each case an effect of the current (for example, the magnetic effect) is used to measure the current (rate of charge flow).

Moving-coil meters can be made very sensitive (i.e. they respond to small changes in current), and when fitted with a **rectifier** they measure **alternating current** as well as **direct current**. They have a uniform scale.

Repulsion-type moving-iron ammeters are not very sensitive or accurate and do not have a uniform scale. They are, however, simple and robust, and measure both ac and dc.

An ammeter goes in series with the element concerned, whereas a voltmeter must be in

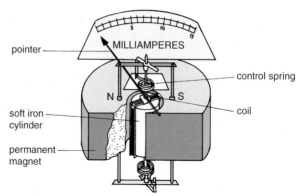

ammeter *A moving-coil meter.*

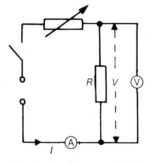

ammeter
Obtaining readings for V, the voltage, and I, the current, allows R, the resistance, to be calculated.

parallel. The circuit shown can be used to measure V/I, which gives a value for R.

ampere (A) The unit of **electric current**. 1 ampere = 1 **coulomb** per second. The unit is named after the French mathematician and physicist André Marie Ampère (1775–1836), noted for his work on electricity and magnetism.

amplifier An electronic circuit whose output **signal** has the same form as the input, but a higher **amplitude**. The ratio of output amplitude to input amplitude is the **gain**. The quality of such circuits is described in terms of gain, freedom from distortion and the range of frequencies they can cope with. A simple **transistor** amplifies, but better output results from more complex circuits.

Many amplifiers work only with ac or other changing input. The dc amplifier is different; it can amplify small constant currents. The **operational amplifier**, or op-amp, is an integrated circuit with a very high gain, used to produce high-quality output.

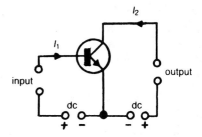

amplifier *A simple transistor amplifying circuit.*

amplitude A measure of the maximum change from the normal or **equilibrium** position in an oscillation (vibration) or a wave; for example, the maximum displacement from the rest position of a pendulum (a), or the maximum value in an alternating current (b). The energy of the oscillation is related to the square of its amplitude: for example, the energy transferred by an alternating current in a resistor is proportional to the square of the amplitude, and the loudness of a sound is proportional to the square of the amplitude of the pressure variations in the sound wave.

analogue or analog A representation of a quantity (usually a changing quantity) that varies continuously in exactly the same way as the original. For example, a simple microphone and amplifier produce an electric current that continuously varies in

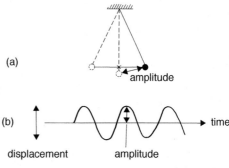

amplitude *(a) A pendulum; (b) a wave.*

proportion to the amount of the air movement in the sound wave at any time. The current may then be used to make a magnetic pattern on tape, the strength of which also varies continuously. Compare this with a **digital** representation, which does not vary continuously but consists of a sequence of numbers that represent the strength of the original at regular intervals. *See diagram.*

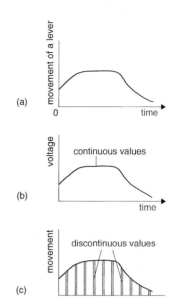

analysis 1. The investigation or testing of substances to find out what they are or what more elementary substances they contain, as in *analytical chemistry.*
   2. The breaking down of a problem into its component parts to help in solving it.
   3. The organization and treatment of data from an experiment to find out what the results mean; for example, by tabulating data, by drawing graphs, and by comparing the data obtained with the predictions or hypothesis being tested.

analogue *(a) original quantity; (b) an analogue copy in a different quantity (voltage rather than distance); (c) original varying quantity sampled. The values of the samples are then digitized (converted to numbers).*

AND gate *See* **gate.**

aneroid barometer A device for measuring atmospheric pressure; 'aneroid' means 'without liquid', as contrasted with the liquid mercury **barometer.** As air pressure changes a springy metal box changes shape. A system of **levers** transfers the changes to a pointer on a scale.

anode The positive terminal of a **direct current** supply, or any positive electrode joined to it. *Compare* **cathode.**

antinode *See* **node.**

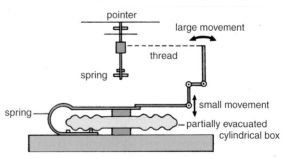

aneroid barometer *A simple aneroid barometer.*

apparent depth The depth that the bottom of a body of water appears to have on the basis of the image formed by light rays from it. Because of the refraction of the light, this is smaller than the true depth: a swimming-pool, for example, looks shallower from the poolside than it really is.

approximation A rough value, one whose **accuracy** is low. It often saves time to work with approximate values rather than with ones of high accuracy.

Archimedes' principle A scientific law that states that the upthrust on a body wholly or partly immersed in a fluid is equal to the weight of the fluid displaced by the body. The *upthrust* (or buoyancy) is an upward force. A *fluid* is a liquid or gas. This law explains why ships and balloons float: their weight is exactly the same as the weight of fluid they displace (the law of **flotation**).

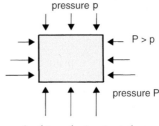

Archimedes' principle

    The upthrust is due to the pressure difference in the fluid between the bottom of the object and its top. Pressure in a fluid increases with depth and so the force due to pressure acting upwards on the base is greater than the downward thrust on the top (*see diagram*).

    The principle is named after the Greek scientist Archimedes (3rd century bc). Many of his theories in physics and mathematics, and his inventions, remain important.

area (A) The measure of the extent of a surface. The scientific unit is the square metre (m^2).

armature 1. The core of soft iron (easily magnetized and demagnetized) on which the wire coil is wound in an electric motor or generator. Its job is to increase the magnetic effect of the current in the coil.

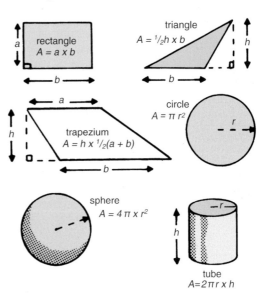

area *Some simple shapes with the equations for their areas.*

2. The soft-iron pole-piece of a permanent magnet or electromagnet; or the core of a solenoid; or the part of an electric bell or magnetic relay attracted by the electromagnet.

astigmatism A defect of lenses (including those of the **eye**) in which the curvature of the surface is different in different directions.

astronomical telescope A **telescope** suited to astronomy. The simplest kind, used by many amateur astronomers, consists of two positive (converging, convex) **lenses**. This produces an inverted image, which does not matter in astronomy. An astronomical telescope should collect as much light as possible, so that faint objects may be seen or photographed. For this reason, the front lens (objective) should have a large diameter. The telescope should also magnify strongly, so that the images of objects such as stars are as widely separated as possible, and the images of objects such as planets and nebulae are as large as possible. To achieve this, the lens near the eye (eyepiece) should have a much greater power than the objective lens. Most telescopes used by professional astronomers are *reflecting telescopes*, in which the objective lens is replaced by a concave mirror.

astronomical unit (AU) The mean distance between the Earth and the Sun. It is used as the basis of distance measurements to nearer stars, using triangulation with the Earth's orbit as the baseline (the method of **parallax**). 1 AU = $1.49597870 \times 10^{11}$ m.

astrophysics The study of the structure of the universe and how the objects in it behave. It includes theories on how the universe, galaxies, stars and planets evolve. Astronomy provides the data used.

atmosphere The layer of gases surrounding a planet, or the outer layers of a star. The atmosphere of the Earth merges into the partial vacuum of space at a height of about 400km, but three quarters of the mass of the atmosphere is below a height of 10 km. Gases

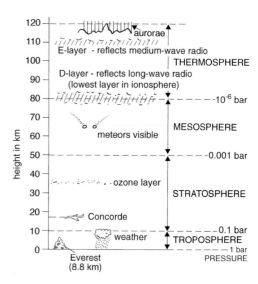

atmosphere *Layers of he atmosphere.*

are kept near the Earth by the force of gravity. The layer is very thin compared to the size of the Earth: if the Earth were an apple, the atmosphere would be no thicker than the apple skin. The gases in the atmosphere are given in the table below, and the main features of the atmosphere are shown in the diagram.

Gas	% volume
nitrogen	78.08
oxygen	20.94
carbon dioxide	0.03
argon	0.093
neon	0.0018
helium	0.0005
krypton, xenon	trace
ozone	0.00006

atmospheric pressure or air pressure The **pressure** at a point in the atmosphere. The force on a square metre at any level equals the weight of air above it. Thus, atmospheric pressure decreases with height. Because of this, simple altimeters (devices that measure height) are forms of **barometer** (devices that measure atmospheric pressure).

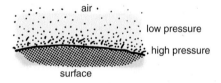

atmospheric pressure *Atmospheric pressure and density (number of particles per unit volume) fall with height.*

The standard value of atmospheric pressure at the Earth's surface is about 100kPa (about 100,000 **pascals**). The actual value varies from place to place and hour to hour, depending on the weather. As atmospheric pressure changes give warning of weather changes, barometers are very useful to forecasters.

atom The smallest particle of an **element** that can have the element's properties. In a simple model of matter, atoms consist of a small massive **nucleus** of **protons** and **neutrons** (**nucleons**) surrounded by orbiting **electrons**. A proton or neutron is nearly 2000 times more massive than an electron.

In normal circumstances, an atom has no electric **charge**; it is neutral, having as many electrons as protons. If it has too few or too many negative electrons, it becomes charged and is called an **ion**.

The diagram below is simplified and is not to scale. Furthermore, the particles cannot be regarded as having definite positions and following definite orbits.

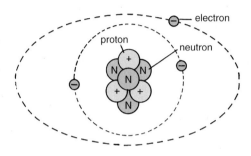

atom *A lithium atom.*

atomic bomb A type of bomb whose destructive force is provided by nuclear **fission** or, in the case of the hydrogen bomb, by nuclear **fusion**. The design of nuclear weapons allows them to release huge amounts of energy in a small space in a short time. This produces a far-reaching blast and a fireball of temperature greater than a million kelvin. These weapons create immense blast and fire damage over very large areas.

As well as releasing huge amounts of energy, the reaction produces damaging radiation, both at the moment of the explosion and from many long-lived radioactive isotopes. These isotopes, present in dust, can travel with the wind for hundreds of kilometres as harmful *fallout*. Fallout may remain a hazard for many decades. *See also* **chain reaction**, **nuclear power**.

atomic mass unit or unified atomic mass constant (u or m_u)
A unit used to express the masses of nuclei and nucleons, equal to one twelfth the mass of a carbon-12 atom.

atomic number *See* **proton number**.

atomic radiations Particles or electromagnetic radiations released in radioactivity, in **fission** or in **fusion**. *See table opposite*.

| | **Type of radiation** | | |
	α (alpha)	β (beta)	ϒ (gamma)
What it is	nucleus of helium atom	high-speed electron	electro-magnetic radiation
Charge	+2	−1	0
Mass in m_u	4	$1/1836$	0
Material needed to absorb most of the radiation	100 mm of air or a piece of paper	2 mm of	Several cm of aluminium lead
Deflection in magnetic field	Slight	Great. In opposite direction to α	None
Ionization that is produced in air	About 2000 ions/mm	About 20 ions/mm	Some ions

audible frequencies Sound **frequencies** from about 20 Hz (**hertz**) to 20,000 Hz, which can be heard by human beings. The higher frequencies in this range become inaudible with old age. Many non-human animals can hear beyond this frequency range. Dogs can hear higher frequencies and can therefore respond to specially designed whistles that give notes above the range of human hearing. *See also* **ultrasonics**.

audio Relating to a frequency in the range 20–20,000 Hz (**hertz**). In the case of sound waves, this is the range the normal human ear can detect (*see* **audible frequencies**). Much electronic hardware handles signals that come from, or become, audible sound waves (for instance, cassette recorders, compact disc players, the telephone system, microphones and speakers). In use, this hardware (*audio equipment*) carries electronic signals in the audio–frequency range, in either **analogue** or **digital** form.

autoclave A vessel that can be tightly sealed so that high temperatures and pressures can be produced inside it. Autoclaves are used to carry out some chemical processes and to sterilize equipment in, for example, surgery. *See also* **pressure cooker**.

average A value that stands fairly for all the values of a set. The usual sense is **mean**; on the other hand, there are cases where the **median** or the **mode** values are wanted. Briefly, the mean value is the result of adding all the values in the set and dividing by the number of values; the median is the value at the middle of the set; and the mode is the most common value in the set.

For example, listed below are the results of eleven groups trying to measure the mass of an object. The results are in milligrams and in order:

749 750 752 752 752 755 755 756 759 762 763

The mean is the sum of those values (8305) divided by 11, that is 755. The median (middle value) is 755 as well, but the mode (most common value) is 752. In practice it is rare for even two of those three types of average to have the same value. The most appropriate type of average must be chosen for each situation; in science, the most common is the mean.

B

back-emf *See* **inductor**.

background radiation Low-intensity **radiation** present in the Earth's surface and in the atmosphere. Radioactive matter exists in nearly all the rocks of the Earth's crust. As this slowly decays, it gives off radiation. All the time too, **cosmic radiation** reaches the surface from the sky. As a result, a **Geiger–Müller tube** gives a count of about one radiation pulse a second.

It is possible that this background radiation may cause thousands of cancer deaths a year. It may also play a part in genetic changes (*mutations*).

So far the background count has not been increased much, if at all, by nuclear weapons testing and power stations. *See also* **radon**.

balance 1. Equilibrium. An object or system is said to be in balance when it is in equilibrium – when the effects of the forces on it oppose each other to give zero acceleration.

2. A device that measures an object's mass or weight by setting up equilibrium between forces. *Beam balances* and *lever balances* are read when there is equilibrium between the opposing **moments** (or *torques*) of the weights of two masses in a gravitational field. As the moments depend on the masses, these machines measure mass rather than weight. The readings would be the same in any non-zero field.

A *spring balance*, on the other hand, measures the object's weight. This opposes the weight with the upthrust caused by the compression of a spring: the reading varies with the gravitational field strength. A spring balance measures weight or force in newtons and is often called a **force meter**, *newton balance* or *newton meter*.

Most balances used in shops, industry and laboratories are now electronic balances, which use electronic circuitry to measure the small changes in resistance in a strain gauge due to the weight of the object being weighed.

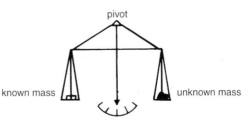

balance *A beam balance.*

balanced forces Forces whose combined effect (**resultant**) is zero. A body will remain at rest, or keep moving at a constant velocity, when acted upon by balanced forces.

band A range of wavelengths of **electromagnetic radiation**, such as the radio-frequency band or **microwave** band.

bar chart *See* **histogram**.

barometer A device used to measure **atmospheric pressure**. In the liquid barometer, there is balance between the air pressure and the pressure due to a column of liquid. This is normally mercury – its high density means that the pressure needed can be reached with a fairly short column. The standard height of the mercury barometer is 760 mm; the changes in the actual value with time help to forecast the weather. However, mercury is costly and toxic and the mercury barometer is relatively large and hard to use. The **aneroid barometer** does not have these problems, but does not give such accurate readings.

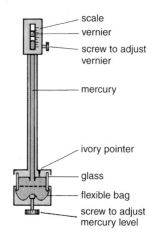

scale
vernier
screw to adjust vernier

mercury

ivory pointer

glass

flexible bag

screw to adjust mercury level

barometer
A mercury barometer.

baryon *See* **particle physics**.

base **1.** A substance able to accept a hydrogen ion in a chemical reaction. Alkalis, such as potassium hydroxide (KOH) and sodium hydroxide (NaOH), are strong soluble bases; they react with acids to produce **salts**. *Compare* **acid**.
2. The control electrode of a **transistor**.

battery A group of single **electric cells**. A car battery consists of six 2–V cells, giving a total voltage of 12 V when in use.

beats An **interference** effect between two waves – for example, between two sound waves at a point in a medium such as air, or between two alternating currents at a point in a circuit. The beat frequency equals the difference between the two original frequencies. *See diagrams opposite.*

becquerel (Bq) The unit used to measure the **activity** of a radioactive source; 1 Bq is an emission of one particle per second. The unit is named after Henri Becquerel (1852–1908), the discoverer of radioactivity.

bel A unit of the power of **radiation**, such as sound or radio, compared to that of a standard. More common is the **decibel**, which is one-tenth of a bel. *See also* **gain**.

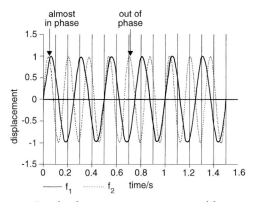

beats *Graph of two interacting waves of frequencies f1 = 4 Hz and f2 = 6 Hz.*

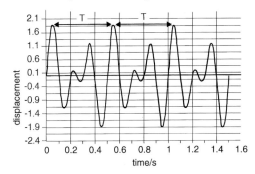

beats *The sum of waves with frequencies f1 = 4 Hz and f2 = 6Hz is a wave with a beat frequency (f2 − f1) of 2Hz, corresponding to a period of 0.5s(T).*

The unit is named after the Scottish-born, American scientist Alexander Graham Bell (1847–1922). He invented the **telephone** and the *phonograph* (an early record player).

bell An electrical device that rings or buzzes as a signal. The electric bell has an **electromagnet** at its centre. When the user closes the switch the iron core becomes a magnet; it attracts the soft iron plate or **armature** and the hammer hits the bell. This movement breaks the circuit as the armature moves away from the screw contact. The current

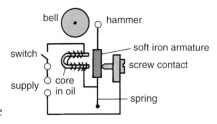

bell *Circuit diagram of an electric bell.*

to the electromagnet is cut off and so the armature is no longer attracted. The spring returns the armature to the position shown, thus closing the circuit again, and the cycle is repeated.

Bernoulli principle The principle that states that the **pressure** in a moving fluid becomes less as the speed rises. For example, if you blow between pieces of paper, they move towards each other. The effect causes the *lift* on an aircraft wing (aerofoil). It is named after the Swiss mathematician and physicist Daniel Bernoulli (1700–82).

beta decay A type of radioactive **decay** in which a **nucleus** emits an **electron**, or **beta (ß) particle**. A **neutron** in the nucleus decays to a **proton** (which stays there) and an **electron** (which escapes with high energy):

$$_0^1 n \rightarrow {}_1^1 p^+ \rightarrow {}_{-1}^0 e + \text{energy} \quad (\,{}_{-1}^0 e \text{ is the } \beta \text{ particle})$$

In the above, superscripts represent **nucleon number**, subscripts represent **proton number**. Thus with the **isotope** $_6^{14}C$, the reaction is

$$_6^{14}C \rightarrow {}_7^{14}N + {}_{-1}^0 e + \text{energy}$$

With this sort of decay the proton number increases by 1. Some radioactive nuclei emit positive electrons (**positrons**, β^+) in positive beta decay. In this case, the proton number decreases by 1, a proton having changed into a neutron. *See also* **alpha decay, gamma radiation**.

The nucleus loses mass and this is converted into energy, which appears as the **kinetic energy** of the beta particle.

beta (β) particle A high-speed **electron** emitted by a **nucleus** during radioactive **beta decay**. Beta radiation is the flow of beta particles from certain active sources, carrying negative charge. Some radioactive nuclei emit positive electrons (**positrons**) in positive beta (β^+) decay. Beta radiation tends to travel farther in matter than **alpha** radiation, but not as far as **gamma radiation**. All three produce **ions** in the matter through which they pass, and thus transfer energy to it. *See also* **radiation, radioactivity**.

Big Bang The period of very rapid expansion (like an explosion) in which the universe was formed, according to current theory (*the standard model*). At the time of the Big Bang, the universe began to exist – creating matter, space and even time. The explosive expansion still continues and astronomers can work backwards from the rate of expansion to estimate that the Big Bang occurred at some time between 10 and 20 billion years ago – measurements are hard to make and have given varied results (*see* **Hubble's law**).

bimetal strip A device made of strips of two metals fixed together. One metal expands more than the other for a given temperature change. As a result, the strip bends by an amount that depends on the change in temperature. The device appears in some **thermometers** and **thermostats**. The diagram shows one in a simple fire alarm; when the temperature reaches a certain value, set by the screw, the strip bends sufficiently to complete the circuit, and the bell starts to ring.

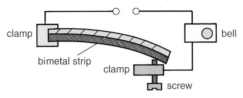

bimetal strip *A simple fire alarm controlled by a bimetal strip.*

bistable Able to exist in either of two states, staying in one of them until a stimulus of some kind is received, when there is a change to the other state. The two states are usually called ON and OFF. An ordinary light-switch is a mechanical example of a bistable device, but the term usually refers to an *electronic* bistable circuit. This usually consists of two logic **gates** linked by a feedback loop – for example, using two NAND gates or two NOR gates. A basic bistable circuit – or simply bistable – can also be made from an OR gate and a mechanical resetting switch. *See diagrams below and on following pages.*

The most common use for a bistable is as a memory element in a computer. Data is stored as binary numbers with ON corresponding to 1 (or logic 1) and OFF corresponding to 0 (logic 0). Bistables are also used in counting, timing and control circuits.

The OR-gate bistable in diagram (a) works as follows (use the **truth table** to help). The OR gate gives an output of 1 when either R or S receives 1, or both of them do. The output at A is fed back to input R. To start with, both inputs S R are at 0, so A is also 0.

Truth table

S	R	A
0	0	0
1	0	1
1	1	1
0	1	1

bistable *(a) Bistable using an OR gate.*

When S is raised to logic 1 output A also goes to 1 – and so, therefore, does R. Now, even if S goes back to 0 the output A stays at 1 because the gate is an OR gate and R is still 1. The output is *latched* at 1. (A bistable device is also called a latch.) To get A back to 0 the reset switch is pressed, which because of the feedback loop also makes R go to 0. Then both inputs are 0 and so is A.

Two NOR gates can achieve the same effect using electronic resetting, as shown in diagram (b). A NOR gate is designed to give an output signal of 1 if *neither* of its inputs is 1. To start with, both inputs R and S are 0. Assume that output A is 1. This situation is clearly self-consistent: NOR1 is receiving inputs of 0 and 1, and therefore outputs 0; NOR2 therefore receives inputs of 0 and 0, and therefore outputs 1, as assumed.

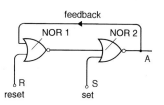

Truth table

S	R	A
0	0	0/1
1	0	0
1	1	0
0	1	1

bistable *(b) Bistable using two NOR gates.*

Putting a signal 1 on input S makes output A go to 0. NOR1's output goes to 1. When S goes back to 0 NOR2's output stays latched at 0. It will change to 1 only if R is made 0, which is done electronically.

Diagram (c) shows a NAND-gate bistable circuit. The details of what happens at each gate are different but the truth table shows the overall action to be the same as in (b).

Truth table

S	R	A
0	0	1
0	1	1
1	0	0
1	1	0/1

(stable in either state)

bistable *(c) Bistable using two NAND gates.*

black body (in thermodynamics) A theoretical object that is a perfect absorber of electromagnetic radiation. Theory shows that such a body would also be a perfect emitter. In practice, a furnace with a small hole in its wall is very close to being a black body. *See also* **temperature radiation**.

black hole The predicted result of the gravitational collapse of a **star** or group of stars. A normal star is at a very high temperature and the particles of matter in it are moving at high speeds. A star has a constant size because there is a balance between the force of gravity due to the star's own mass, which tends to compress it, and the tendency for the high-speed particles to escape from the star. At the end of its life, a star loses the ability to convert its mass into energy and the star cools. Ultimately, if the star is massive enough, gravity will compress it to such a size that its gravitational field is so strong that not even light can escape from it. Any nearby matter will be pulled into the star. As it cannot emit light and swallows up such matter it is called a 'black hole'.

Stars with masses at least three times greater than the Sun should eventually collapse to form black holes. Black holes are impossible to see and difficult to detect, but when matter falls into them it loses so much energy that it emits X-rays, which can be detected. A small number of likely black holes have been detected in this way.

block diagram A simplified way of showing a **circuit** by giving only the main functions of groups of elements (components). *See diagram* at **radio**.

boiling temperature or boiling point The temperature at which the vapour pressure of molecules escaping from a liquid surface becomes equal to the external pressure. Vapour bubbles form inside the liquid. The normal boiling temperature of a pure liquid is defined as the boiling temperature when the external pressure is the standard air pressure, of approximately 100 kPa. Thus the normal boiling temperature of pure water is 100 °C at that pressure.

The boiling temperature falls if the air pressure falls – at the top of a high mountain, for example. The boiling temperature rises if an impurity such as salt is added. *See also* **evaporation**.

Boiling	Evaporation
Occurs at one temperature – the boiling temperature at that pessure	Can occur at any temperature, but increases as temperature rises
Occurs within the liquid	Occurs only at the surface of the liquid
Bubbles appear	No bubbles

bond energy The energy required to pull atoms free from each other's influence and therefore the energy holding atoms together.

Bourdon gauge An instrument that measures gas or liquid pressure by means of a coiled metal tube which unwinds when pressure increases. The movement is transferred to a pointer which moves against a calibrated dial.

Boyle's law One of the three **gas laws**, which states that for a constant mass of gas at constant temperature, the product of pressure and volume is constant. This applies to any **ideal gas** – that is, a gas at a temperature well above boiling. If a gas sample at constant temperature goes from one state (p_1, V_1) to a second (p_2, V_2) then $p_1V_1 = p_2V_2$. This means that if the pressure doubles, the volume halves, and so on. Pressure and volume are inversely proportional.

The law is named after Robert Boyle (1627–91), one of the founder members of the **Royal Society**.

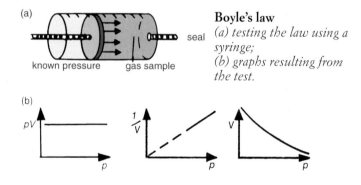

Boyle's law
(a) testing the law using a syringe;
(b) graphs resulting from the test.

breezes **Convection** currents in the air. Land and sea breezes occur near the coast; they result from the fact that the temperature of the land rises and falls more quickly than that of water (see **specific thermal capacity**). Thus, on a hot day, (a) (see diagram), warm air over the warm land rises; cool air from the sea comes in to replace it. On a cool night the reverse occurs as the sea is warmer than the land, (b).

Brownian motion The random movement of tiny particles – specks of dirt, for

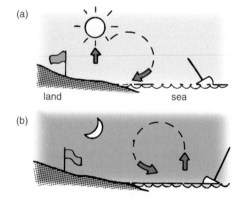

breezes (a) A warm day produces a sea breeze; (b) a cool night produces an offshore breeze.

example – in a fluid. Each speck is so small that the impacts of fluid particles from different directions do not balance. Brownian motion is good evidence for the **kinetic model** of matter. The movement of the specks, which can be seen, gives evidence of the movement of particles which cannot be seen by the naked eye.

Brownian motion is named after Robert Brown (1773– 1858), a Scottish doctor who travelled to Australia soon after it was settled and brought back specimens of 4000 new plant species. He discovered 'Brownian motion' in 1827 when he was working on pollen grains. The same year he became head of the British Museum's botany department.

Brownian motion *The continuous bombardment of a particle by invisibly small particles of a fluid causes random movement.*

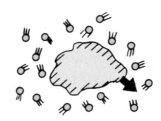

bubble chamber A detector for **fundamental particles**. It contains a liquid, often hydrogen, under pressure, kept just beneath its **boiling temperature**. The pressure in the chamber is decreased just before subatomic particles pass through. Any of the particles that are charged leave trails of charged hydrogen atoms, or **ions**. The hydrogen begins to boil, bubbles of hydrogen gas appearing along the trails of ions. The trails are photographed and the pressure is again increased to cause the hydrogen bubbles to liquefy again. The bubble chamber has largely superseded the **cloud chamber**.

buoyancy The effect of the upward force (**upthrust**) on an object in a fluid. An object is said to be buoyant if it is hard to sink. *See also* **Archimedes' principle**, **flotation**.

C

cable rating The maximum safe current that a cable or wire is designed to carry. Electric wires or cables must be thick enough to allow the current for which they are designed to pass without causing overheating. The rating is given as a measure of the maximum safe current (usually at 240 volts in Britain). Typical values are in the range from 3A (for a lamp) to 68A (for a cooker). *See also* **mains plug**.

calibration The process of putting a usable and reliable scale on the indicator of any measuring instrument. For example, in the calibration of a mercury-in-glass thermometer two fixed points (the **melting temperature** of ice and the **boiling temperature** of water, say) are marked, and then the distance between the mercury levels at each temperature is graduated into the hundred divisions of the Celsius scale.

Calorie (Cal) or kilocalorie A unit of energy content of food. It is being superseded by the kilojoule: $4.2kJ = 1$ Cal. The calorie (small *c*) has been replaced in scientific usage by the joule: 1 cal $= 4.2J$.

camera An **optical instrument** able to produce a permanent record of a scene by focusing light onto a light-sensitive material or device – for example, a film treated with light-sensitive chemicals. In the photographic camera shown, the film absorbs focused light from the scene; the chemical changes produced depend upon the brightness and colour of the light absorbed at each spot. Processing the film makes the chemical changes permanent.

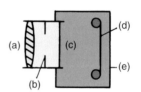

camera *The parts of a camera: (a) lens; (b) diaphragm; (c) shutter; (d) film; (e) camera body.*

   The operations necessary for a camera to take a photograph are as follows:
(i)   The distance between **lens** (a) and film (d) is set so that the **image** is sharply in focus;
(ii) The *iris diaphragm* (b) is set at an opening (aperture) that lets the right amount of light through;
(iii) The *shutter* (c) opens, allowing light onto the film for the right length of time.
These parts are in a light-tight box, the camera *body* (e).

camera, pinhole A **camera** with no **lens** (*see diagram*). Without a lens to give a focused image, only a very small hole (*aperture*) can be used. Otherwise the image is blurred. Because the small hole lets through little light, the film in a pinhole camera needs a long exposure. It can be used, therefore, only for stationary objects.

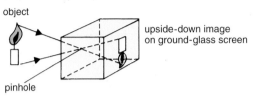

camera, pinhole

capacitance A measure of the quantity of charge that an object can hold. Capacitance is measured in farads. An object with a capacitance of 1 farad holds 1 coulomb of charge when it is at an electric potential of 1 volt. The defining equation (capacitance in farads = charge in coulombs/ potential in volts) is:

$$C = \frac{Q}{V}$$

See also **capacitor**.

capacitor A device that stores electric charge. A typical capacitor used in electric and electronic circuits consists of a pair of metal plates separated by a **dielectric**. Capacitors are charged by connecting them to a power supply via a resistor (*see diagram* (a)). To start with, when the charge on the capacitor is zero there is a large current in the resistor (*see graph, diagram* (b)). As charge builds up so does the voltage across the capacitor, and the current falls. When the voltage across the capacitor equals the supply voltage, current becomes zero and the capacitor is fully charged. More charge will flow if the supply voltage is increased.

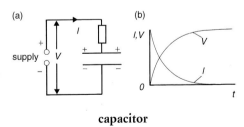

capacitor

The **capacitance**, C, of a capacitor will increase when
• the area of the plates is increased
• the plates are moved closer together
• a dielectric is chosen that has a higher *dielectric constant*.
Variable capacitors are often used in tuning circuits in radios. The

capacitance is varied by changing the area of overlap of the plates. Capacitors are used in changing ac to dc (*see* **rectifier**), and in many electronic applications. As charge stores they keep the clocks in computers working when not connected to a power supply, and charged capacitors provide the large but short-lived currents for camera flash units.

capillarity The tendency of a liquid to rise or, sometimes, to fall in a narrow tube. It is an effect of **surface tension**. A liquid usually rises up a narrow tube, pulled upward by the liquid's concave **meniscus**, as in diagram (a). Mercury, however, will sink down such a tube because its meniscus is curved downwards, as in (b). *See also* **adhesion**.

A *capillary tube*, as used in mercury thermometers, is one with a very narrow bore.

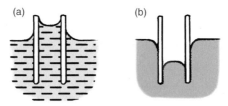

capillarity Capillary action in (a) water and (b) mercury.

cathode The negative terminal of a **direct current** supply or any negative electrode joined to it. It is the electrode that becomes plated in electrolysis by metals whose positive ions are attracted to it. *Compare* **anode**.

cathode rays An old-fashioned term for the stream of electrons emitted from the negative electrode (**cathode**) in a vacuum tube (such as a TV tube). The term was used before it was discovered that cathode rays did in fact consist of electrons.

Originally cathode rays were emitted simply by placing a high voltage between cathode and anode (*cold emission*), but modern devices use an **electron gun** in which the electrons are emitted from a heated filament (**thermionic emission**). Such electron streams are controlled by electric and/or magnetic fields to produce images on phosphorescent screens in TV tubes, computer monitors, radar screens, electron microscopes etc. *See also* **cathode-ray oscilloscope, cathode-ray tube**.

cathode-ray oscilloscope (CRO) A device used to make measurements and display data using a narrowly focused beam of electrons, which produces a spot of light when it strikes a phosphorescent screen. The beam is controlled by electric fields between the X plates, which move the beam horizontally, and between the Y plates, which move the beam vertically. The vertical displacement of the beam (and so of the spot on the screen) can be used to measure the voltage applied between terminals connected (usually via an amplifier) to the Y plates. As electrons have a very small mass the beam can be moved very quickly. This allows the CRO to be used to follow rapidly changing voltages, like those produced in radio and audio applications. These usually involve waves of some kind, and are displayed by using a timebase, a circuit that applies a steadily increasing voltage to the X plates, and so draws the spot horizontally across the tube at a constant rate.

cathode-ray tube (CRT) A visual display device used in **television**, **radar**, **cathode-ray oscilloscopes** and some computer screens. In each case a fine beam of high- energy electrons leaves an **electron gun** and makes a spot of light where it meets the screen. The colour of the light produced depends on the nature of the coating of the screen. Electric and/or magnetic fields in the **vacuum** through which the beam passes deflect the beam and therefore the spot.

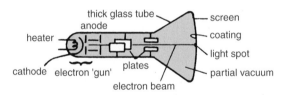

cathode ray tube *Cross-section of a CRT.*

Although complex and bulky, the CRT is fairly cheap, because so many are made. Portable computers (laptops) use screens made of semiconductors. These are expensive but are completely flat and compact. Large flat screens are hard to make, but may eventually replace the CRT.

cell, electric A device in which an **electromotive force** (voltage) is produced by two different **electrodes** (usually metals, alloys or carbon) separated by a chemical solution (**electrolyte**). When the electrodes are connected by an external circuit, charge flows around the circuit, gaining energy to do so from chemical changes in the cell. The maximum voltage produced by a cell depends on the chemicals used, but is usually between 0.5 V and 2 V. Cells may be connected in series to make a **battery**, which provides a higher voltage. For example, the simple carbon–zinc cell (*see diagram overleaf*) produces 1.5 V; a 6-V battery can be made using four of these in series. *Primary cells* like the carbon–zinc cell use up the chemicals in irreversible reactions and have then to be discarded. *Secondary cells* are rechargeable; the chemical changes that produce the emf are reversible. An example is the car battery, which consists of six lead–sulphuric acid 2-V cells to provide the standard 12 V. Many portable devices (such as laptop computers and cassette players) now use rechargeable cells, which have to be much lighter than lead–acid cells. Much research in **electrochemistry** is now taking place to produce light, large-capacity rechargeable cells.

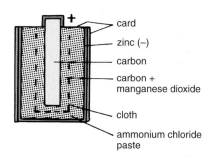

card
zinc (–)
carbon
carbon +
manganese dioxide
cloth
ammonium chloride
paste

cell *A 1.5V carbon–zinc dry cell.*

Celsius, Anders (1701–44) A Swedish astronomer who directed the Uppsala observatory, making it one of the most important in the world. He is best known, however, for devising a centigrade scale of temperature – one with 100 degrees between the ice and steam temperatures ('grade' is an old word for 'degree'). In fact, the first Celsius scale had 100° for the ice temperature and 0° for the steam temperature. Soon after Celsius' death, one of his students turned it the other way up to make it what we now use.

centre of gravity (cg) The point through which the **weight** of a body can be thought of as acting. Since the force of gravity is effectively constant over the small volume of an object on Earth, cg and **centre of mass** are for practical purposes the same point. The cg of a flat shape can be found by hanging it, in turn, from several points near the edge. Each time, the cg is under the point of suspension. Vertical lines drawn from the points of suspension with the aid of a plumb-line cross at the cg. *See also* **stability**.

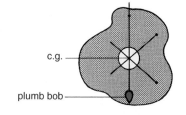

c.g.

plumb bob

centre of gravity

centre of mass The point in an object or system at which, for many purposes, its mass can be thought of as being concentrated. Thus the Sun, planets and moons are all point masses as far as their motions and the forces between them are concerned.

The centre of mass of a simple object of uniform density is at its geometric centre. It is marked c in each of the cases shown in the diagram.

  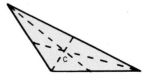

centre of mass *The centres of mass (c) of simple geometric shapes.*

centrifugal force An apparent or fictitious force directed away from the centre for an object moving in a circle. A force directed towards the centre is required to keep an object moving in a circle (*see* **centripetal force**), and if, for example, you were in a car taking a corner at speed you would feel this central force as the car body pressing against you. However, you interpret this as a (fictitious) force pushing you into the car body. It is best to avoid the term centrifugal force when discussing circular motion.

centripetal force The centrally directed force required to keep an object moving in a circle – that is, to provide the acceleration that is directed towards the centre of the circle. An object will move in a straight line at a constant speed unless there is a net (unbalanced) force acting on it. When this force acts at right angles to the direction of motion, the object will accelerate in the direction of the force and move in a circle. The force acts at right angles to the tangent of the circle, and so is always directed towards its centre. When a car turns a corner by moving in a circular arc the centripetal force is provided by road friction. Gravity provides the centripetal force keeping satellites and planets moving in their orbits.

The diagram shows an object moving in a circle at the end of string. The tension in the string provides the centripetal force. If the string breaks, the object will not move outwards but will keep moving in a straight line, which is a tangent to the circle.

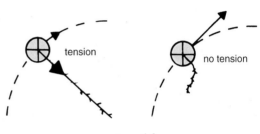

centripetal force

Chadwick, James (1891–1974) A British nuclear physicist, best remembered for his discovery of the **neutron** in 1932. For this he gained a Nobel Prize in physics.

chain reaction A process in which one event causes a second of the same kind, which leads to a third, and so on. An important example is in nuclear **fission**. Among the products of the fission (splitting) of a nucleus of ^{235}U (*see* **uranium**) are **neutrons**. If a neutron of the right energy hits a second ^{235}U nucleus, it can cause that to split in turn. More neutrons result, and so the chain reaction continues.

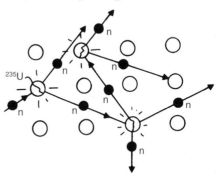

If, on average, the neutrons from a fission cause more than one new fission, there is a 'runaway' chain reaction. This is what happens in an **atomic bomb**. To harness controlled **nuclear power**, the reactor is designed so that, on average, each fission causes exactly one further fission.

chain reaction *The chain reaction resulting from the fission of 235U.*

charge *or* **electric charge** A property of some **fundamental** particles that gives rise to forces of attraction or repulsion between them. There are two kinds of charge, named *positive* and *negative*. **Protons** carry a positive charge, **electrons** a negative charge. The fundamental law of **electrostatics** states that like charges repel, unlike charges attract.

Normal matter contains equal quantities of positive and negative charge, so the net effect is that matter in bulk is neutral or uncharged. In most materials, electrons are easily removed from their atoms. This can often be done by rubbing against another material; the material that loses electrons becomes positively charged, while the one that gains electrons becomes negatively charged (*see* **static electricity**). **Atoms**, **molecules** or groups of atoms (as free radicals) may gain or lose electrons and so become *charged particles* – **ions**. Charged particles in motion make up an electric **current**.

The unit of charge is the **coulomb** (C). It is the quantity of charge transferred by a current of 1 **ampere** in 1 second. The basic *elementary charge* is that carried by an electron or a proton: 1.6×10^{-19} C. A coulomb is equivalent to about 6×10^{18} elementary charges.

charge distribution The way electric charges are arranged on the surface of a conductor. If an object has a net **charge**, and charges can move in it, the forces between them will make them move apart. The net charge of the sample will therefore appear on the surface. For complex reasons, the more curved the surface, the greater the charge density there. Thus a sphere will have a constant charge density, any charge in a cube will appear only on the edges, and nearly all the charge on a pointed object will gather at its tip.

Because of this last effect, the charge density at a point can be so high that there is charge transfer to or from outside. This leakage can **discharge** a charged object; the pointed tip of a **lightning** rod reduces the chance of a stroke of lightning.

charging The gaining of net electrical **charge** by a neutral object. There are several methods of charging an object. If the surface of the object is rubbed with a suitable second surface, it can gain or lose **electrons**, to become negative or positive. This is charging by friction. The object can also gain charge by contact with a charged surface. Charging by induction involves applying an electric field to the object, so that positive and negative charges separate. If the object is then **earthed**, charge of one sign will flow away from the object, leaving it with the opposite kind of charge.

In *photoemission*, electrons are released from a surface by radiation (*see* **photoelectric effects**). The object becomes positively charged unless more electrons enter from elsewhere.

Charging a battery (*see* **cell**) means reversing the normal chemical changes of discharge. Work is done by an external device (for example, a **generator**) to increase the electrical potential energy of the ionic chemicals in the battery.

Charles' law One of the three **gas laws**; it states that for a constant mass of an **ideal gas** at constant pressure, the volume is proportional to the absolute **temperature**. This means that if the sample temperature T doubles, so will the volume V: $V_1/T_1 = V_2/T_2$.

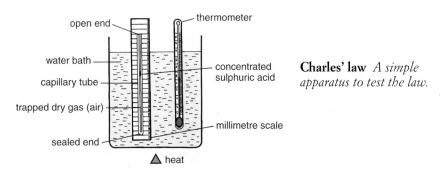

Charles' law *A simple apparatus to test the law.*

The law is named after Jacques Charles (1746–1823), professor of physics at Paris, where he studied how gases behave. He made the first hydrogen balloon ascent for this purpose.

chart A graphic way of showing information. Maps and block diagrams (for instance, electric circuit diagrams, evolutionary trees and experiment equipment sketches) are charts, and so are **graphs**: they show how aspects of some systems relate to each other. In fact, well-designed charts are often able to do this with much more effect than a few paragraphs of text.

chip *See* **integrated circuit.**

circuit A number of electrical elements (components) joined by conductors to a source of electrical energy. *Circuit diagrams* show in simple form the order in which the components are connected. (*See Appendix D for usual symbols.*) In the circuit shown in diagram (a), the source supplies current (through switch and fuse) to a lamp. The meters measure the current in the lamp and the voltage across it. In practice, diagram (b) may be more like the real thing. However, each component has many types, and circuit diagrams are much easier to draw, print and use. Sometimes complex circuits are simplified into **block diagrams**.

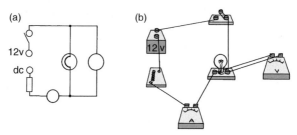

circuit *(a) A circuit diagram;*
(b) how the circuit looks in reality.

circuit breaker A safety switch that automatically cuts off the **current** when there is an overload. It is often used in place of a **fuse** nowadays, as the cut-off current can be set more precisely. A *residual-current circuit-breaker* (RCCB) should be used with devices like electric lawn mowers when a dangerous leak to earth may still not be enough to blow a normal type of fuse.

circular motion The motion of a body in a circle. Even if the body moves at a constant speed, its velocity changes continuously because the direction of motion, and therefore the **velocity**, is changing – that is, the

body is accelerating. It can be shown that the
acceleration is directed towards the centre of the
circle. This means that a force must act on the body
towards the centre of the circle – a **centripetal force**.
For example, for a satellite in a circular orbit, the
centripetal force is provided by gravity; for a mass
whirling on the end of a string, the centripetal force is

circular motion

provided by the tension in the string. The force (F) increases with the mass
(m) of the object and its speed (v). The force decreases as the radius (r)
increases. The formula is:

$$F = \frac{mv^2}{r}$$

classical physics The **physics** of a hundred or more years ago. It is the
body of knowledge developed before **relativity** and **quantum physics**, and
before the discoveries of **cathode rays, X-rays** and **radioactivity**. *Modern
physics* includes all these fields, as well as others such as **electronics** and
photoelectricity (*see* **photoelectric effects**).

climate The long-term pattern of weather in a region. Britain's climate is
temperate (that is, it never becomes very hot or cold); it is modified by it
being adjacent to the Atlantic Ocean and bathed by the warm Gulf Stream.
   The balance between energy input from the Sun and energy reflected or
re-emitted from the Earth and the atmosphere is a crucial influence on
climate. *See also* **global warming, greenhouse effect**.

clock signal A regular sequence of electric pulses used to synchronize the
passage of data in a computer. The higher the *clock rate*, or rate of the
clock signal, the faster the computer can handle data.

cloud chamber A detector for **fundamental particles**. Air in the
chamber is just saturated with vapour. Ionizing radiation passing through it
leaves a trail of **ions**. These act as centres for the vapour to condense; a line

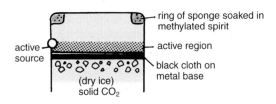

cloud chamber *A diffusion cloud chamber.*

of droplets appears. The diagram shows a **diffusion** cloud chamber; the vapour density is always suitable for droplet formation in the shaded region. The cloud chamber has been largely superseded by the **bubble chamber**.

cohesion The force of attraction between particles (usually molecules) of the same substance. Cohesive forces cause the effects of **surface tension** and are involved in **evaporation**. The attractive force between particles of different substances is called **adhesion**.

collector One of the terminals of a **transistor**.

colour The sensation of different wavelengths of visible light. Colour is sensed by different sets of cone-shaped cells in the retina of the **eye**, one for each primary colour, blue, green and red. *See* **colour vision**.

The wavelength range of visible light is from about 400 nm (deepest red) to about 760 nm (deepest blue). This range contains several hundred *hues* (distinguishable colours), which merge into each other. The range is often divided into six or seven bands, as in the table (*right*). White does not appear here – it contains all hues equally.

l/nm	Colour
400–420	violet
420–450	indigo
450-500	blue
500–560	green
560–600	yellow
600–650	orange
650–760	red

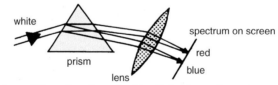

colour *The refraction of white light through a prism.*

Refraction of a parallel beam of white light by a glass prism can produce the **spectrum** of colours as shown in the diagram (*below*). A second method uses **diffraction**. The process is called **dispersion** in each case.

Note that in refraction red light bends less than blue.

Few surfaces reflect all light. Most absorb some wavelengths more than others. As a result only certain colours are reflected, and the surface seems 'coloured'. 'Blue' paint largely absorbs all colours except blues and greens, which it reflects strongly. In red light it will look darker than in white light, but not completely black, since it is likely to reflect a little red light.

'Yellow' paint may reflect red, orange and green light, as well as yellow; it will look dark in blue light.

Most paints (*pigments*) are mixtures. Blue paint mixed with yellow looks green, as only greens are reflected. This *subtractive colour mixing* differs from the additive effects of mixing coloured lights.

In *additive colour mixing*, as in colour television, the three *primary colours* are red, blue and green. An equal mixture of these will give white light.

Additive secondary colours are formed by any two of the three primary colours:

$$yellow = red + green \text{ (or minus blue)}$$
$$magenta = red + blue \text{ (or minus green)}$$
$$cyan = green + blue \text{ (or minus red)}.$$

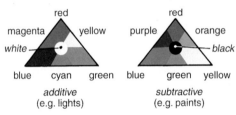

| additive | subtractive |
| (e.g. lights) | (e.g. paints) |

colour *The effects of colour mixing.*

colour code A system of colours painted on to **resistors** and **capacitors** to denote the value of the resistance in ohms or the capacitance in picofarads. Resistors and capacitors are too small to have their values and tolerances written on them clearly, but the colour code is now being replaced by a digital code with simpler markings as shown in the table overleaf.

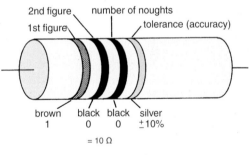

colour code
The colour coding of a resistor.

Figure	Colour	Figure	Colour
0	black	5	green
1	brown	6	blue
2	red	7	violet
3	orange	8	grey
4	yellow	9	white

colour code *Table of colour code values.*

Value	0.27	1	3.3	10	220	1k	68k	100k	1M	6.8M
Mark	R27	1R0	3R3	10R	K22	1K0	66K	M10	1M0	6M8

colour code *Table of replacement digital codes.*

colour vision The ability of people and some other animals to distinguish between light wavelengths of different **colours**. The process starts in the cone cells of the retina of the **eye**. People's eyes differ in the range of light wavelengths they can detect, and some people are *colour-blind* – that is, they cannot clearly tell the difference between certain colours.

comet A small body made of ice, grains and dust, which orbits the Sun, usually in a highly elliptical path. The grains are small lumps of rock surrounded by a layer of water ice and frozen gases. Solar radiation melts the ice and frozen gases, and pushes particles away from the Sun to form a brightly illuminated tail. Most comets are believed to originate in a huge belt of such bodies far beyond the orbit of the farthest planet (Pluto), called the *Oort Cloud*. Some come from a belt of small objects called the *Kuiper Belt*, lying closer in, just outside the orbit of Neptune.

communication The transfer of information (knowledge) between people. Modern communications systems depend on electronics and are an important part of **information technology**; examples include the phone network (the biggest machine in the world), radio and television; other important systems are **fax**, **satellite** communications, electronic mail, the World Wide Web and the internet.

compass A small **magnet** used to show the direction of the Earth's magnetic field. The box is placed level; the user turns it until the ends of the needle point to N and S on the card. It is then easy to find the other directions. The north pole of a compass points to the north of the Earth because the Earth behaves as though there is a magnetic south pole in the region of geographic north. *See also* **magnetic material**, **Earth's magnetism**.

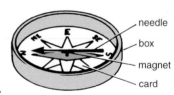

needle
box
magnet
card

compass

component 1. Any part of a structure, system or object, such as the lamp or the switch in a lighting circuit.
   2. One of the two or more elements that may be obtained by resolving a **vector** in different directions. Resolving a vector into components is an aid to analysis. For example, a ball thrown at an angle to the horizontal has a

vertical component and a horizontal component of velocity, and treating them separately makes it easier to predict the path followed by the ball (*see diagram*). Components combine to form the **resultant**, which simply reproduces the original vector. *See also* the **parallelogram rule**.

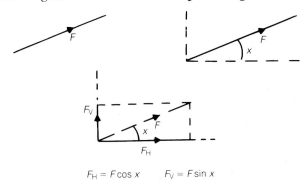

$$F_H = F \cos x \qquad F_V = F \sin x$$

component *Steps in finding the components of a vector F.*

composite material An artificial structural material that combines the properties of more than one substance; for example, glass-reinforced plastic (GRP), concrete, and steel reinforced with carbon fibres.

compression Force or pressure applied to a body in such a way as to make it smaller or shorter.

conclusion The final, and most important, result of an **experiment**. An experiment's aim should be to answer a clear question about the world, and the conclusion is the answer to that question. In some cases, experimenters may conclude that they cannot answer the question, or that the first idea they had is wrong; experiments like that are at least as important in practice as those with a positive outcome.

conduction, electrical The process of **charge** transfer through a medium. A *conductor* is a substance able to pass charge with ease; a **semiconductor** is fairly good; *insulators* allow charge transfer only when the applied **voltage** is very high. A voltage applied to a sample involves making one end positive compared to the other. If the sample contains charges which are free to move, the voltage forces them towards the oppositely charged end.

Charge passes easily through metals. This is because a metal contains many (negative) **electrons** which are not bound to other particles, so are free to move. The voltage makes these electrons move. As they move, they collide with each other and with the positive cores that form the bulk of the metal. The collisions slow the electron motion, creating **resistance** to the current. The energy transfer causes a rise in temperature.

The **current** (rate of flow of charge per second) depends on the voltage and on the resistance. *See* **Ohm's law.**

Note that it is conventional to define current direction as being from positive to negative. In many cases, as in the diagram, the current in fact consists of negative charge moving the other way.

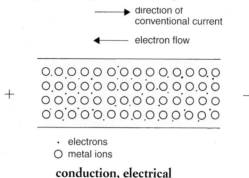

· electrons
O metal ions

conduction, electrical

Some liquids conduct. They contain negative and positive **ions** and are called *electrolytes*. Current passing through them has chemical effects. *See* **electrolysis.**

Gases are insulators. However, at low pressure a fairly low voltage gives a small current. Energy is released in the form of radiation such as light; this **discharge** through gases is used in 'neon' lamps (which can use various gases as well as neon) and fluorescent lamps. Here too the process is one of ion transfer.

A **vacuum** is an insulator. But if the voltage is high enough, electrons are forced out of a negative electrode in a vacuum and move at high speed through empty space. *See* **cathode rays.**

State	Medium	Conduction mode
Solid	Metal	Electrons
	Semiconductor	Electrons and/or holes
Liquid	Metal	Electrons
	Electrolyte	Ions
Gas		Ions and electrons

conduction, thermal The transfer of **internal energy** from particle to particle in matter. *Compare* **convection**, **radiation** and **evaporation.** Substances differ in their ability to conduct in this way. Non-metals are very poor conductors, and are used for thermal **insulation**. Metals are good thermal conductors; they contain a 'gas' of free-moving **electrons**, which carry energy quickly from points at high temperature.

conservation of energy *See* **energy, law of conservation of**.

constellation One of the 88 named areas of the sky used to identify and name stars and other heavenly bodies. They were originally based on certain star patterns which seemed to resemble animals, gods, heroines and heroes in Greek or Roman mythology – for example, *Taurus* (the Bull), *Ursa Major* (the Great Bear), *Orion* (the Hunter), *Andromeda* (the heroine rescued by the hero *Perseus*). The stars in a constellation are labelled with Greek letters, generally, though not always, in order of brightness: the brightest star in a constellation is labelled α (alpha), the next brightest β (beta), and so on. This is followed by a modified form of the Latin constellation name. So the brightest star in Orion is named α Orionis, but, like many bright stars, it also has its old name, *Betelgeuse*. Most star names are Arabic in origin, as the Arabs produced the first detailed star catalogues.

control 1. A parallel experiment carried out at the same time as a main experiment with one feature altered so that the results can be compared and the extent of any change quantified. For instance, in an experiment to find the effect of adding a substance to a chemical reaction, the control would be the reaction without the substance.

2. The field of automating and managing processes using machines (including the use of robots). A *control system* accepts inputs from people and **sensor**s, processes these on the basis of a set of stored instructions, and outputs actions. A good control system has a clear feedback path for data between output and input; in other words, sensors measure the output to keep it within set limits. The diagram shows a simple control system, aimed at keeping the thickness of rolled sheet metal within the desired range.

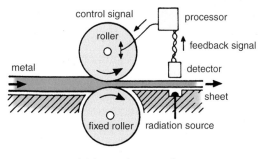

control *A simple control system.*

convection A thermal process in which energy is transferred by the bulk movement of a heated fluid. When a fluid (for example, air or water) is heated, it expands and becomes less dense than surrounding, unheated, fluid. The less dense material floats upwards (*see* **flotation**, **Archimedes' principle**) and is replaced by denser, cooler fluid. This is heated and, in turn, floats upwards, so producing a *convection current* in the fluid (*see diagram*). Note that according to the **kinetic model** (or theory), the fluid expands because its particles have gained kinetic energy, so move faster and increase the pressure they exert on surrounding, cooler fluid. *Compare* **conduction**, **evaporation**, **thermal radiation**.

Winds and **breezes** are caused by convection currents in the atmosphere.

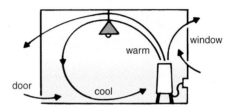

convection *How a convector heater heats a room.*

cooling curve The graph of the temperature of a substance as it cools down over a period of time. It is particularly useful for finding the **melting temperature** of a substance – the temperature at which it solidifies. *See* **latent heat**. Another use is to find energy losses due to cooling.

Copernicus, Nicolaus (1473–1543) A Polish mathematician and doctor who revived and developed an ancient theory, the Sun-centred model of the universe. Before that time, most people believed the Earth was fixed at the centre of the universe. The change in view required new physical knowledge (provided by **Galileo**), and a lower estimate of the importance of human beings. The change was therefore not an easy one, but it was a major part of the development of modern science at the end of the European dark age.

coplanar In the same *plane* (two-dimensional surface).

cosmic background radiation The radiation that fills the universe, of very nearly equal intensity in all directions. It is at the very low temperature of 2.7 K, and is one of the main pieces of evidence in favour of the **Big Bang** model of the origin of the universe. In this model, the universe began at a very high temperature (10 billion K one second after the Big Bang) and has since cooled down to 2.7 K.

cosmic radiation Very high-energy particles from space. Their source is not certain – they could come from stars that erupt as *novae* or explode as *supernovae*. Cosmic particles are mostly protons or nuclei of light atoms; they reach the Earth from all directions at very high speeds. When such a particle collides with a particle in the air, a *shower* starts. By the time this reaches the surface of the Earth, it may cover hundreds of square kilometres. *See also* **background radiation**.

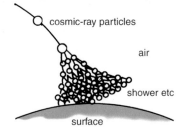

cosmic-ray particles

air

shower etc

surface

cosmic radiation
A shower of cosmic particles.

cosmology The study of the structure and evolution of the universe. The current main theory in cosmology, the *standard model*, links the observed expansion of the universe with **particle physics** and the **Big Bang**. *See also* **Hubble's law**.

coulomb (C) The unit of electric **charge**; a coulomb is the charge transferred by a current of 1 ampere in 1 second. It is named after Charles Coulomb (1736–1806), a French engineer whose main work concerned electricity and magnetism.

counter An electronic circuit able to count pulses of **current** even at high rates. A **Geiger–Müller tube** is one source of such pulses; it may also drive a speaker. *Compare* **scaler**.
   The number of pulses counted in unit time (seconds or minutes) is the *count rate*.

couple A pair of equal and opposite parallel forces tending to turn an object. The **moment** or torque (turning effect) on the object is the product of one force and the distance between them.

couple *The torque (T) produced by a couple. F1, F2 are equal, opposing forces; s is the distance between the forces.*

F_1

s

$T = F_1 s \ (\text{or } F_2 s)$

F_2

critical angle *See* **total internal reflection**.

critical temperature The temperature above which it is impossible to compress a gas into a liquid. The critical temperature of water is 374°C. *See also* **vapour**.

crystal A piece of matter whose atoms are arranged in a regular structure. Crystals are normally solid, but it is possible for some liquids to show crystalline structure – *see* **liquid crystal**.

The crystals of a given substance tend to have the same shape. The shape of a perfect crystal is determined by the structure of the atomic array. The shapes of actual crystals are modified by the conditions of growth of the crystal. Crystal structure adds to the evidence for the **kinetic model** of matter.

Curie A family of French physicists who did much early research on **radioactivity**. The best-known is Marie (1867–1934), born in Poland, whose discovery of radium has been of major value. She shared the 1903 Nobel Prize for physics with Henri Becquerel; her second prize, for chemistry, was in 1911. Her husband, Pierre, and daughter, Irène, also did much useful work in physics. The latter, with her husband, discovered fission in 1934, without realizing what they had found. Both died of radiation-induced cancer.

The family name lives on in a unit of radioactivity, the curie; the element curium is an **alpha-particle** emitter widely used in power sources. *See also* **Curie temperature**, named after Pierre, who first worked on it.

Curie temperature The temperature above which a magnetic material cannot be magnetized. The Curie temperature of pure iron is 770° C. Above this value the energy of the particles is too great for them to hold together in **domains**.

current (I) The rate of flow of **charge** (*see* **conduction, electrical**). The unit is the **ampere**, symbol A. 1 A is a current of 1 **coulomb** per second, $I = Q/t$. *See also* **Ohm's law**.

Current is usually measured by its effects. The ampere is defined in terms of its magnetic effect. Other effects of an electric current are the heating and electrochemical effects. The use and control of currents in circuits are of immense importance in everyday life, communications, transport and industry.

current balance A device used to measure **current**. The user must balance weights against the magnetic force between coils carrying the current (*see diagram*). The **ampere**, the unit of current, is defined in terms of the force

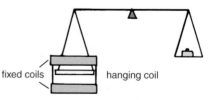

fixed coils   hanging coil

current balance

between wires that carry current. The current balance can therefore measure current *absolutely*.

cycle A single member of a recurring series of changes in, for instance, a **wave**, a **vibration**, or a rotation. At the end of a cycle, the system has returned to its state at the start. *See also* **hertz**.

A common cyclic change is the sine wave; a single cycle is the sine wave corresponding to one circle of 360°.

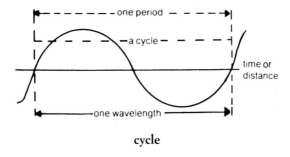

cycle

D

Dalton, John (1766–1844) A British scientist whose most important work was on the physics and chemistry of gases. He was red/green colour-blind and described his experiments on this form of colour-blindness, which is still sometimes called Daltonism.

data 1. A series of observations, measurements or facts.
2. Information inside a computer memory or other information system. A *databank* (or *database*) is a large collection of data arranged for people to use.

data logging The process of collecting **data** from the environment automatically with a *data logger*. This uses **sensors** to capture the data and feed it into a computer system; the system stores it for later processing (**analysis**) and display as instructed.

data storage The holding of **data** in some electronic form that allows easy access to it. Many systems use magnetic material in the same way as an audio tape recorder: magnetic tapes and discs are in wide use with computers of all sizes. **Integrated circuits** and compact discs are of growing importance too; they can hold a lot of data in easy-to-use form.

day and night The periods of light and darkness brought about by the rotation of the Earth. The Earth is a sphere which turns on its axis in slightly less than 24 hours. At a given time, part of the Earth faces the Sun, giving daylight to that part, and part faces away and is in darkness. Thus while it is midday in Britain, it is evening in India and morning in America, and the middle of the night in New Zealand. *See also* **seasons**.

decay A change of a property such that its value decreases with time. In the study of **radioactivity** it describes how the activity of a substance changes with time.

A graph like the one shown appears in either context. For example, the measure on the y-axis may be the number of given active nuclei in a sample, the charge on a **capacitor**, or the **amplitude** of a damped **vibration**.

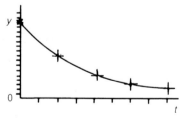

decay *A graph showing exponential decay.*

Such a graph shows an *exponential* decay, meaning that the value of y halves during a certain period of time. *See also* **half-life**.

decibel (dB) The unit used to measure the intensity of a sound, or of radio or electrical signals; 1dB is one tenth of a **bel**. The decibel scale measures the *ratios* of intensities. The size of the unit is chosen so that an increase of 10 dB produces a tenfold increase in intensity, for example, the 'loudness' of a sound. A sound too quiet to be heard is rated at 0 dB, quiet speech is at 60 dB and a sound that is loud enough to cause pain is at about 130 dB. Note that the 'loudness' of a sound is not quite the same thing as intensity, since the loudness of a sound depends on the properties of the ear and brain, not just of the energy in the sound.

demagnetization The removal or loss of magnetism. A **magnet** can lose its magnetic strength when treated roughly or stored without **keepers**. To demagnetize a specimen completely, slowly withdraw it from a powerful ac **solenoid** to a distance of a metre or so along the axis.

density (ρ) The mass of any substance divided by its volume. The SI unit is the kilogram per cubic metre, kg/m^3, although the gram per cubic centimetre, g/cm^3 is often used.

$$\text{density } \rho = \frac{\text{mass } m}{\text{volume } V}$$

The density of water is 1000 kg/m^3 (1 g/cm^3). It is often more helpful to use *relative density*, D. This is the substance's density ρ_s divided by that of water. So $D = \rho_s/\rho_w$. There is no unit.

The density of iron is 7900 kg/m^3. Its relative density is therefore 7.9.

As solid and liquid volumes change little with temperature and pressure, their densities are fairly constant. On the other hand, the density of a gas varies greatly with temperature and pressure. *See also* **optical density**.

dependent variable A variable whose value changes when the value of another quantity or factor is changed. Often an **experiment** explores how changing one thing affects the value of a second. Both are variables, measures able to have different values – the one that is altered is the *independent variable*; the one that changes as a result is the dependent variable (it depends on the first). As a general rule, the independent variable is plotted along the *x*-axis, and the dependent variable along the *y*-axis, but this rule is not always followed.

detection of radiation Making visible, recording, or measuring radiations that may be invisible or undetectable by human senses. Nuclear radiation can be detected in many ways, such as with a **Geiger–Müller tube** or a **cloud chamber**. The method of detecting electromagnetic radiations depends on the type. For instance **X-rays** darken photographic film, **infrared** (thermal) rays affect the current in a phototransistor and **ultraviolet** rays eject electrons from freshly cleaned zinc.

deuterium   The hydrogen **isotope** of **nucleon number** 2 (rather than the normal value of 1). The deuterium **nucleus** ('*deuteron*') contains a neutron as well as a proton. Deuterium is thus twice as dense as normal hydrogen; many of its physical properties are not the same, as these depend on density. Deuterium is often called 'heavy hydrogen'.

Heavy water is deuterium oxide, D_2O. Its melting and boiling temperatures are 4° C and 102° C; its density is more than 10% greater than that of H_2O.

dielectric   An electric insulator in which charge separates (polarizes) in an electric field and may remain separated when the field is removed. Dielectrics are used to make **electrets**; these have oppositely charged faces. *Compare* **magnet**. *See also* **insulation**, **polarization**.

Good insulators of this type are used between the plates of **capacitors**. *Relative permittivity (dielectric constant)* measures their value in this context.

diffraction   The bending of a wave round the edge of an opaque object, into the shadow region. It is easy to see the diffraction of water waves in a **ripple tank**.

diffraction   *The diffraction of water waves seen in a ripple tank.*

All waves can diffract round suitable objects. The effect is large only if the size of the object is roughly the same as the wavelength of the waves.

Light diffracted by a narrow slit produces a set of **interference** fringes. The **shadow** of a hair shows fringes at the edges and a bright line in the centre.

As diffraction depends on wavelength, it can lead to **dispersion**; the *diffraction grating* is a useful source of spectra. Because of interference each wavelength can leave in only a few directions.

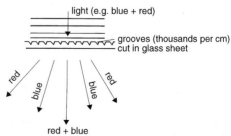

diffraction *Action of a diffraction grating.*

diffuse reflection **Reflection** by a rough surface, such as the paper of this book – and indeed by most visible objects. A flat, smooth surface like that of a mirror reflects a parallel beam as a parallel beam. A rough surface does not, so cannot produce **images** (*see diagram*). Nevertheless, the laws of reflection are obeyed.

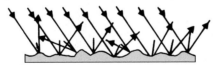

diffuse reflection *Light waves reflected from a rough surface are not parallel and so cannot produce an image.*

diffusion The spreading of one substance through another by the random motion of particles. The particles spread from a region of high concentration to regions of lower concentrations. This is the way a scent will spread through a room, or an ink drop spread throughout a beaker of water. Diffusion can also occur, if more slowly, in solids and gels. *See also* **kinetic model**.

digital Relating to digits. *See* **digitization**.

digital display A display using the digits 0 to 9 (in a decimal scale) to show a reading, rather than a scale like that of a thermometer or the pointer of an **analogue** meter.

digitization A process in which the value of a quantity is coded as a number (or *digit*), usually a binary number. The process is carried out electronically. Digital coding is now widely used in information (data) storage and transmission, such as in compact discs and some telephone and broadcast systems. The number is stored or transferred as a series of pulses, giving the on–off of the binary 0–1 logic system. It is usually coupled with a sampling process which converts a continuously varying quantity into a set of whole numbers that follow the changing values of the quantity. The advantage of digital coding is that the series of pulses is less likely to suffer from distortion or noise than an **analogue** equivalent.

diode A device that allows charge to flow through it in one direction only.

A diode has two electrodes; a modern diode is usually a **semiconductor** *p–n junction*. Its main use is as a **rectifier**, as illustrated in the diagram. When an alternating signal voltage is applied to

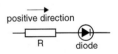

diode

the diode–resistor combination, the voltage across the resistor becomes positive only, since no current flows in the negative direction: the signal has been rectified.

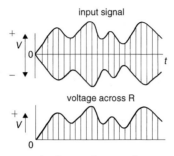

input signal

voltage across R

diode *How a diode rectifies an alternating signal.*

dioptre (D) The unit of the **power** of a **lens**, **mirror** or system to deflect light. It is the reciprocal of the **focal length** in metres. Thus, a 5-D lens has a focal distance of 0.2 m:

$$\frac{1}{0.2} = 5$$

direct current (dc) A flow of **charge** in one direction only. *Compare* **alternating current (ac)**. In most cases the current can be expected to be fairly steady. Thus the current shown in the diagram would be taken as a direct current with an ac ripple.

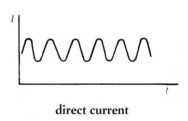

direct current

discharge The transfer of **charge** from a charged object. The word is often used in the special cases of the discharge of a chemical source or a **capacitor**, and the current in a gas. In the latter case a high voltage is needed to give sparking. *See* **conduction, electrical**.

Discharge tubes include neon and fluorescent lamps. These contain gases at low pressure; a fairly low voltage causes current, with energy emitted as radiation. Often much of the radiation is **ultraviolet**. To change this to visible light, the tube is coated with a **fluorescent** powder.

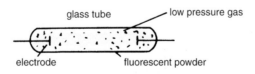

glass tube   low pressure gas

electrode   fluorescent powder

discharge *A fluorescent lamp (tube).*

dish aerial A device that collects electromagnetic radiation by using a concave reflecting surface to focus the rays onto a detector. Dish aerials are widely used to collect TV signals from satellites, but originated in **radar** and **radio astronomy**.

dispersion The separation of radiation according to frequency, as in the production of a spectrum by a prism. When radiation bends, because of **refraction** or **diffraction**, the angle of bending depends on frequency. For instance, when white light passes at an angle from air into glass, blues bend more than reds. The white light is dispersed (spread out). *See also* **colour**.

displacement The **vector** that describes an object's change of position. The unit is the metre, m; the symbol is s. The **scalar** form is distance. In the diagram the displacement of C from B is l, 4 km. If an object moves from A to B to C, the distance moved is 7 km; its displacement is 5 km in direction 054°.

displacement *The displacement s from A is 5km in the direction 054°.*

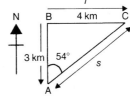

In **wave** motion, displacement measures the disturbance from zero at a given point at a given time. The largest value is the wave's **amplitude**.

distance ratio or **velocity ratio** The distance moved by the input force (or effort) acting on a machine divided by that moved by the output (or load) in the same time. There is no unit. *See also* **efficiency**.

distortion *See* **amplifier**.

domain A region with a uniform magnetic field in a **magnetic material**. The particles of a magnetic substance have fields as a result of the flow of **charge** in their electron clouds. In a domain all the particle fields are in the same direction.

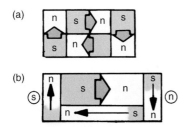

domain *Domain fields in (a) an unmagnetized sample and (b) a magnetized sample.*

In an unmagnetized magnetic sample, shown in diagram (a), the domain fields point in different directions; there is no net effect. When the sample is magnetized, some domains grow, and their magnetic fields increase in strength, to give a net field, as in (b).

Doppler effect The apparent frequency change of waves when the source is moving with respect to the observer. Thus a police car siren has a higher pitch than normal during approach and a lower pitch as it moves away. The **red shift** of light from stars has the same kind of cause.

The effect is named after Christian Doppler (1803–53), an Austrian physicist who was a professor in Prague and then in Vienna.

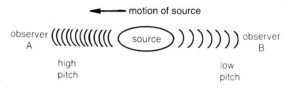

Doppler effect

double insulation An electrical safety arrangement in which any live wires in a device are insulated by at least two layers of non-conducting material. This is an alternative to using a third earth connection – the green and yellow wire in a three-wire mains cable. This means that the device uses just two wires to connect to the mains supply, the live and the neutral. *See also* **earth**.

drag The resistive force produced on a vehicle as it moves through the air.

dynamics The study of the motion of objects and the forces causing changes in the motion. *See also* **equations of motion, Newton's laws of motion**.

dynamo *See* **generator**.

E

ear The sense organ of hearing, able to detect **sound** waves of frequencies between about 20Hz and 20kHz (in humans). The incoming vibrations move the eardrum, (a), to and fro. This motion is passed through the middle ear by a series of small bones, (b). In the inner ear is the cochlea, (c), a coiled tube of liquid. From this the auditory nerve, (d), passes signals to the brain; these relate to the frequency and loudness of the sound input. *See also* **audible frequencies**.

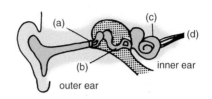

ear *The components of the human ear: (a) eardrum; (b) small bones; (c) cochlea; (d) auditory nerve.*

earth To join an object to the Earth by a conductor so that it will share any net **charge** it has with the Earth. As the Earth is so large, this means in effect that an earthed object cannot keep a charge. In practice, something is earthed by being connected to a metal plate in the ground, as in diagram (a).

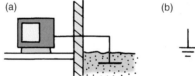

earth *(a) An object is earthed by connecting it to a metal plate in the ground; (b) the standard symbol for an earth lead.*

The green and yellow striped wire in British household cable should be joined to Earth; this will reduce danger if there is a fault (*see* **mains plug**). The earth connection is a protection for users. Without it a person could receive a fatal electric shock; the fuse would not guarantee protection, since the current might not increase sufficiently to blow the fuse.

Earth The third planet out from the Sun in the **solar system**, after Mercury and Venus. The Earth is nearly a perfect sphere of rock, almost 13,000km across; its mass is around 6×10^{21} tonnes. Its average distance from the Sun is about 150 million kilometres, and it takes a year to go round the Sun once (*see* **seasons**). While the Earth moves in orbit round the Sun, it turns on its axis once a day (just under 24 hours) to produce the cycle of **day and night**.

It seems that the Earth is the only planet with any kind of life forms. These exist in a very wide range of **environments** in a very thin layer: the lower **atmosphere**, all parts of the land surface, and the seas and oceans to the greatest depths.

The outer crust of the Earth is 5–65km thick, being thinnest under the oceans and thickest under the continents. Most earthquakes and volcanic activity originate here. The mantle is the denser rock beneath the crust, extending about halfway towards the centre. Its temperature increases with increasing depth. The core consists of iron with a little nickel. Its outer part is molten, while its central part is crushed into the solid state by the huge pressure there.

The behaviour of air and water vapour in the lower atmosphere causes the weather. The upper atmosphere (from about 60km above sea level) is important for radio communications and some **satellites**. *See also* **ozone layer.**

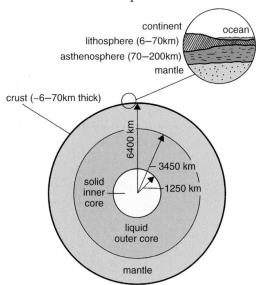

Earth *The structure of the Earth. The lithosphere comprises oceanic and continentaltectonic plates. These move over the asthenosphere, which is softer and partly molten.*

earth leakage circuit breaker A residual current **circuit breaker.**

earthquake The event that produces shock waves (**seismic waves**) due to a sudden slippage of rocks in the Earth's crust. The sudden movement relieves the stress on rocks caused, usually, by the movement of tectonic plates. The effects of earthquakes are measured using the *Richter scale*, which is related to the energy released in the earthquake. Each successive point on the scale represents an increase in the energy released by a factor of approximately 31.6. *See also* **plate tectonics.**

earthquake waves *See* **seismic waves.**

Earth's magnetism The magnetic properties of the Earth. The Earth has a **magnetic field** that extends many thousands of kilometres above the surface. The cause may be current inside the Earth, mainly in the nickel-iron core. The field is like that of a bar **magnet** with an S-pole under the Arctic.

echo The reflection of a sound or other kind of wave from an object or surface. Echoes are used in sound- ranging or **sonar** as shown in the diagram. Sonar is used to locate submarines and fish. It is also used by bats, dolphins and other animals to locate their prey. The sound from explosive charges is used in geological research to find the boundaries between different rock layers, and in oil exploration. **Radar** echoes are used to locate aircraft or ground objects using short radio waves (microwaves), and is the method used to measure planetary distances and so the **astronomical unit**. The time between emission of the wave and the return of the echo is measured, and to analyse the results the speed of the wave in the media involved must be known.

echo *An echo sounder is used to measure the depth of water beneath a ship.*

eclipse The passing of one astronomical body behind the other. For example, one star in a pair of gravitationally linked stars moves behind the other (*see* **eclipsing binary**); the Sun is eclipsed when the Moon passes in *front* of it, as viewed from the Earth. In all eclipses one body cuts off light emitted from another. An eclipse of the Moon occurs when it passes into the shadow of the Earth. In this case, we do not see the eclipsed body passing behind another, but being cast into shadow. The diagrams show how eclipses of Earth and Moon are produced. Total eclipses occur in the *umbra*; partial eclipses in the *penumbra*.

eclipsing binary A pair of stars circling each other whose orbits are such that each star in turn moves in front of the other in relation to the Earth. The result is a periodic change in the brightness of the star system. Their importance lies

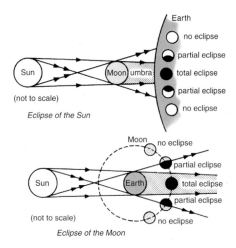

eclipse

in the fact that Newton's laws can be used to calculate the masses of the stars by timing the brightness variation.

eddy current A **current** that is induced (*see* **electromagnetic induction**) in a conductor when it is in a changing magnetic field. The effect can result in the loss of useful energy – for this reason, the cores of electric machines are laminated, i.e. made of thin strips, so that large currents cannot pass. Eddy currents can be used to slow down the motion of a conductor, for instance in eddy-current braking and in the speed-meters of cars. This effect is called *electromagnetic damping.*

efficiency The ratio of useful work done, or energy output, to the work or energy input, in an energy transfer system, such as a machine or an engine. The efficiency ratio is usually given as a percentage. The missing energy is not destroyed but is either used to do non-useful work (such as lifting pulleys) or is lost as heat to the surroundings. For example, an electric light bulb converts only about 2% of the input energy into light – the rest heats the surroundings. The laws of thermodynamics limit the efficiency of **heat engines** and power stations. A car engine has an efficiency of about 25%.

The efficiency of a simple **machine** may be calculated as

$$\frac{\text{work output}}{\text{work input}} \times 100\% \quad \text{or} \quad \frac{\text{force ratio}}{\text{distance ratio}} \times 100\%$$

Einstein, Albert (1879–1955) A German-born American physicist, best known for his work on **relativity** (1905 and 1915). He worked in many other fields as well, notably **quantum physics**. Einstein was a patent-agent when he published his first major works in 1905. His first wife, Mileva Maric, was a physicist too; she is thought to have helped him in his work a great deal in those early days. Einstein gained the 1927 Nobel Prize for his work on photoelectricity.

elasticity The property possessed by an object or substance of returning to its original shape and size after the removal of a force that has deformed it. Many, but not all, elastic materials also obey **Hooke's law**. Objects or substances that are permanently deformed by a force show *plasticity* and are called **plastic**. For example, copper wire extends elastically and obeys Hooke's law for small forces, but behaves plastically at larger forces.

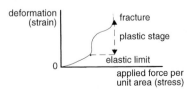

elasticity *The stress/strain curve of metal sample.*

Metals tend to behave elastically up to a certain value of **stress** (measured as force per unit area) called the *elastic limit*, after which they suffer permanent deformation.

electret A permanently polarized **dielectric** material. Its electric field is similar to the magnetic field of a permanent **magnet** – that is, it has a positive charge at one end and a negative charge at the other. Small microphones use electrets that move as a result of the sound wave and induce currents, forming an electrical signal.

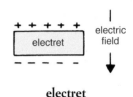

electret

electric field A region of space where there are electric forces on any **charges** present. It is common to describe electric fields using *lines of electric force*, or *field lines*. Such a line is defined as the path a free positive charge would take. The lines show the *field direction* by their direction, and the field strength by their closeness.

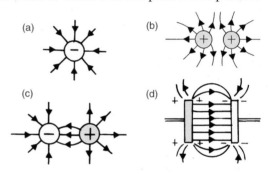

electricity The nature and effect of electric **charge** – whether in motion (**current**) or stationary (**static electricity**). The movement of electric charge involves the expenditure or input of **energy**, and electricity is a major energy source.

electric field *The fields around (a) a lone charge; (b) two like charges; (c) two unlike charges; (d) two charged plates.*

electricity distribution The transfer of electrical energy to users from power stations. The process involves the use of step-up **transformers** to increase the voltage of the supply from the 25,000V produced by **turbines** to much higher voltages (275,000V and 400,000V) used by the national grid system. This increase in voltage means that smaller currents can be used to transmit the power demanded, with a considerable decrease in the energy losses in the cables, which are proportional to the square of the current. The current is carried by well-designed multicore cables, either underground or (mostly) above ground on tall pylons. The high transmission voltages are reduced by step-down transformers at the users' end. *See also* **electricity generation**.

electricity generation The production of electricity for the needs of society. The normal way to do this in highly populated countries is to build numbers of large power stations. These produce electricity by burning a fossil fuel, or by using the **energy** of nuclear fission or falling water. Other methods being explored include the use of solar energy, tides, and wind. *See also* **electricity distribution, hydroelectricity, nuclear power**.

It is also possible to generate electricity on a small local scale. This is common for remote farms and villages, ships and spacecraft. *See also* **generator**.

electricity in the home Electric power enters the home from a main cable outside, passing through the main switch, meter and **fuse**/switch box. In most houses, the cable forms a ring main, with the lamps and sockets linked to that as required. Fuses and **circuit breakers** protect the wiring (and the user) from overload; the meter records the energy used.

Electric current has various effects; some examples of their uses are:

(a) *Heating effect*: fire, cooker, toaster, iron, filament lamp (made hot enough to emit light).

(b) *Motor effect*: motors in video machine, vacuum cleaner, fridge, washing machine, analogue clock, heating pump.

(c) *Magnetic effect*: bell, buzzer.

(d) *Other uses* of energy: power for electronic circuits (for example, audio, TV), phone, digital clock, computer, microwave oven.

electrochemical series *See* **reactivity series**.

electrochemistry The study of chemical changes and reactions that involve electricity. In theory this covers practically the whole of chemistry, but in practice the main topics are as follows:

• the *electrochemical series* of metals, in which metals are ranked according to their reactivity

• the chemistry of ionic substances (acids, bases, salts)

• **electrolysis**, in which electric currents produce chemical effects, notably electroplating

• electric **cells** (batteries) used to make portable electricity supplies.

electrode The conductor by which electric charge leaves or enters a conducting substance, in particular an electrolytic cell, a battery or the low-pressure gas in a 'vacuum' tube such as a TV tube. In use, the positively charged electrode is the **anode**, the negative is the **cathode**. *See also* **electrolysis**.

electrolysis The production of chemical changes by passing an electric current through an *electrolyte*. The changes rely on the fact that an electrolyte, either a molten ionic substance or one dissolved in water, contains **ions** that move to form the current. The ions involved are **discharged** at the **electrodes**, positive ions moving to the **cathode**, negative ions to the **anode**.

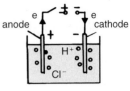

For example, in a solution of hydrochloric acid the significant ions are positive hydrogen ions (H^+) and negative chlorine ions (Cl^-). At the anode, chlorine ions lose electrons and combine to form neutral molecules of chlorine gas, which bubbles off:

electrolysis
The electrolysis of HCl.

$$Cl^- + Cl^- \rightarrow Cl_2(g) + 2e^-$$

At the cathode, hydrogen ions gain electrons and form neutral molecules of hydrogen, which also bubbles off:

$$H^+ + H^+ + e^- \rightarrow H_2(g)$$

In effect, electrons have passed through the electrolyte as they have through the rest of the circuit.

Electrolysis is used in many applications in industrial chemistry. It is an aid in the manufacture of chemicals (for example, aluminium), and in the electroplating of strong, cheap metals with a thin layer of another more expensive metal, for protection against corrosion or to improve appearance. It is also the process by which secondary **cells** are recharged.

electrolyte *See* **electrolysis**.

electromagnet A device that acts as a **magnet** when a current flows through its coil. The core is usually made of iron. The strength of the effect depends on the number of turns in the coil, the nature of the core, and the **current**. *See also* **bell**, **relay**, **speaker**.

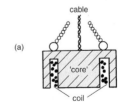

electromagnet damping *See* **eddy current**.

electromagnetic induction The generation of a voltage (**electromotive**

electromagnet (*a*) *An electromagnet;* (*b*) *the poles of an electromagnet viewed end-on – the polarity is determined by the current direction.*

force) across a conductor when this moves with respect to a magnetic field. This induced electromotive force is larger when there is:
• a higher relative motion between the conductor and the magnetic field;
• a greater length of conductor in the field (for example, more turns on a coil);
• a stronger magnetic field.
Thus, in the test shown in diagram (a), fast relative motion gives a high reading of V; slow relative motion gives a low reading (but for a longer time); no motion gives zero voltage.

Lenz's law states that the direction of the induced voltage is such as to generate a current whose magnetism tends to oppose the original magnetic field. There is no emf unless the conductor cuts through **lines of force**.

To produce the voltage, charge must flow in the conductor. The direction of the voltage (and thus of the charge flow) is as shown in diagram (b) when an S-pole moves towards the coil; it is the same when an N-pole moves away from it. In both cases the direction of the current gives an S-pole at the left end of the coil (and an N-pole at the other); this opposes the change that caused it.

Electromagnetic induction is the basis of many electric machines, such as the **generator** and the **transformer. Eddy currents** and back-emf are problems caused by the effect. *See* **right-hand rules**.

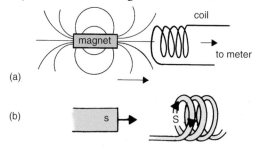

(a)

(b)

electromagnetic induction (*a*) *Demonstration of electromagnetic induction; (b) Lenz's law ensures an S-pole at the left end of the coil.*

electromagnetic radiation A range of **radiations** that differ in wavelength but have all the following properties:
(a) They are caused by moving electric charge.
(b) They consist of vibrating electric and magnetic fields.
(c) They pass through empty space at 300,000,000 m/s, the **speed of light**.
(d) They tend to be absorbed by matter, in which they travel more slowly.
(e) Like all **waves**, they show **reflection, refraction, interference** and **diffraction** effects.

(f) They are transverse waves and show **polarization** effects.
(g) They can show particle properties.
For practical reasons, the **spectrum** of electromagnetic waves is split into a number of regions.

Region	Wavelength	Frequency (Hz)
gamma	-10^{-12}	10^{21}
X-ray	10^{-12}	$10^{18}-10^{21}$
ultraviolet	$10^{-12}-10^{-7}$	$10^{15}-10^{18}$
visible light	$10^{-7}-10^{-6}$	$\approx 10^{15}$
infrared	$10^{-6}-10^{-3}$	$10^{7}-10^{12}$
microwave	$10^{-3}-10$	$10^{7}-10^{12}$
radio	$10-10^{6}$	$10^{2}-10^{7}$

electromagnetic waves *Electromagnetic wave regions. Note – all figures are approximate and some regions overlap.*

electromagnetism The study of the interaction between an electric current and a magnetic field, or the magnetic effects produced. *See the preceding entries.*

electromotive force (emf) (E) The voltage (or potential difference) produced by a source of electricity that can drive a current through a circuit. Forces in an electrical source are able to move **charge** against the circuit resistance R to form a **current**. The source also has resistance r to the flow of charge through it, called the **source resistance**. The forces may be due to the separation of charged particles (in a **cell**) or to electromagnetic effects (in a **generator**).

Electromotive force, often called *voltage*, is measured in terms of the energy involved in the transfer of one **coulomb** of charge. The unit, the volt, is the joule per coulomb (1 V = 1 J/C).

Emf (E) is larger than the available (external) voltage, as source resistance r has to be allowed for in most circuits: $E = I(R + r)$.

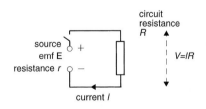

electromotive force $E = I(R+ r)$.

electron One of the basic particles of matter. Its mass is around one two-thousandth that of a **nucleon**; it carries a negative **charge**. There are electrons in all normal **atoms**; they form a cloud around the nucleus.

In a **metal**, electrons act rather like the particles of a gas; this *electron gas* gives metals their shiny surface and makes them good conductors, both thermal and electrical. *See also* **cathode ray; conduction, electrical; emission**.

electron gun The arrangement of electrodes that accelerates electrons ('cathode rays') in a **cathode-ray tube** or TV tube.

electron (vacuum) tube A glass tube that contains low-pressure gas and into which **electrodes** protrude. Until the development of **semiconductor** systems, it was the basis of electronics. One electrode is hot, so can give out electrons (*see* **emission, electron**); these pass through the tube in a way that depends on the number, structure and voltages of the other electrodes. The only such tube in common use now is the **cathode-ray tube**.

electronics The study of the control and applications of the motion of electrons in **metals**, **semiconductors**, **gases** and empty space. *Electronic engineering* deals, on the whole, with much smaller currents than electrical engineering; its circuits tend to be smaller and involve a wider range of elements with no moving parts. Its branches include audio, video, communications systems and computer systems.

Microelectronics concerns the design and use of **integrated circuits**.

electroscope A device that allows the user to detect whether an object is charged, and if so, with what sign. Modern versions, *electrometers*, also allow the size of the **charge** to be found.

The leaf electroscope is still very common in teaching. The central part is a thin, light piece of metal foil fixed to a metal rod. Charge given to the cap spreads through the metal. The leaf rises as it has the same charge as the rod and so is repelled by it. In normal use, the case is earthed (*see* **earth**). The angle of the leaf indicates the voltage between case and rod.

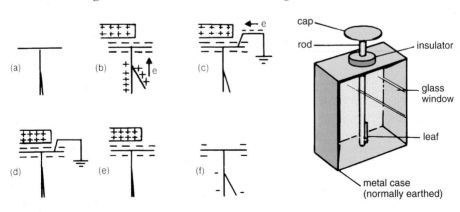

electroscope *Charging an electroscope by induction.*   **electroscope**

electrostatics The branch of physics concerned with **static electricity**: that is, with electric charges that do not move because they are on insulators like polythene or acetate. *See also* **dielectric**.

element 1. A substance that cannot be split chemically into simpler components. An element is composed of **atoms** all of which have identical chemical properties and behaviour. The atoms of an element may differ in the number of **neutrons** in their nuclei (*see* **isotope**).
 2. The high-resistance coil in an electric heater or kettle.

elementary particle An old term for **fundamental particle**.

emf *See* **electromotive force**.

emission, electron The escape of **electrons** from a surface, usually a metal. Some of the electrons in a metal are free to move internally but are still bound to the metal. They can escape if they acquire enough kinetic energy (*compare* **surface tension, evaporation**). The different types of electron emission gain this energy in various ways:
(a) *thermionic emission*: energy is supplied by heating the metal, thus raising the internal kinetic energy of the particles. This method is used in most **electron tubes**, especially **cathode-ray tubes**.
(b) *photoemission*: energy is supplied by illuminating the surface with electromagnetic radiation. The radiation here acts as a stream of particles (**photons**). This process usually needs high-energy (high-frequency) **ultraviolet radiation**. *See also* **photoelectric effects**.
(c) *secondary emission*: energy is supplied by the kinetic energy of particles (electrons, ions) which bombard the surface.
(d) *field emission* or *cold emission*: energy is supplied via a strong electric field set up between the emitting surface and a nearby positively charged surface.

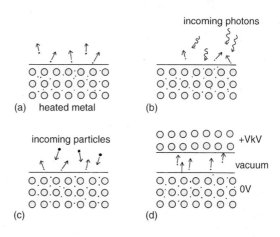

emission, electron *The different types of electron emission: (a) thermionic emission; (b) photoemission; (c) secondary emission; (d) field emission.*

emission spectrum *See* **spectrum**.

endoscope A device used mainly in medicine to make observations inside the body. It consists of a thin tube containing many even thinner optical fibres, and is inserted into the body via, for example, a blood vessel or the alimentary canal.

energy (W) The ability of a *system* to do **work**. Work is done when a force moves something. A system consists of at least two elements, for example:

- a mixture of a fuel and oxygen, which can do work via an engine when combustion takes place
- a mass at a height above the Earth's surface, which can do work via a machine when the mass is allowed to move downwards under the force of gravity
- a moving mass which can do work via a machine when slowing down relative to another object

The unit for both energy and work is the **joule** (J).

It is sometimes helpful to think of different forms of energy, although a more modern approach prefers to think of ways in which energy is transferred.

The fundamental forms of energy are few. **Potential energy** is the energy of a system in which an object is in a field that exerts a force on it. The amount of potential energy is equal to the energy that would be released if the body moved to some reference position. Thus a gravity field (g) exerts a force on the mass of the object, giving *gravitational potential energy* (mgh), where h is the height of the mass above the reference level.

An electric field acts on the charge of the body, if any, and gives rise to *electrical potential energy*, which again is defined relative to some reference position.

Kinetic energy ($1/2mv^2$) is the energy possessed by a mass (m) moving relative to the observer with a velocity (v); it is measured by the work the object could do by being brought to rest (that is, to the same speed as the observer – usually at rest on the Earth).

Radiation energy is the energy possessed by electromagnetic waves, which are a combination of moving electric and magnetic fields. The radiation can do work by, for example, acting on charged particles inside a substance, giving them both kinetic and potential energy, thus heating the substance.

Other forms of energy are in general one of these three or some combination of them. For example,

- *chemical energy* is simply a shorthand way of describing the electric potential energy which binds atoms together and may be allowed to do work as a result of chemical reactions
- *heat* or *thermal energy* is the combination of electrical potential and kinetic energy of vibrating particles (in solids and liquids) or the more freely moving particles in a gas. This is often called *internal energy*
- *nuclear* or *atomic energy* is the potential energy that arises from the strong forces that bind **nucleons** together in a **nucleus**.

It is sounder and often more useful to think about energy transfer processes. Energy can be transferred by doing work and by **heating**.

Electrical transfer is the process that uses the motion of charged particles to do work. For example, the forces working on electrons in a resistor push them against other electrons, atoms and ions, and give them extra kinetic and potential energy. This raises the internal energy and thus the temperature of the resistor, i.e. it heats it. The energy transferred is equal to the quantity of **charge** flowing through the resistor multiplied by the voltage across it (QV). In terms of time t and current I, the energy transferred is VIt, and results in a gain in internal energy ('heat') of $mc\Delta\theta$ where m is the mass of the resistor (or system being heated), c its **specific thermal capacity** and $\Delta\theta$ the rise in temperature.

Thermal transfer is the transfer of energy from one object to another (heating and cooling) via energy transfers involving **conduction**, **convection**, **evaporation** or **thermal radiation**. These are essentially random processes, because they involve the random motion of particles in the substances, whereas doing work is an ordered process.

Energy is not an easy concept to understand, and it is often simpler to think about and describe the processes by which work is done and by which energy is transferred, rather than worry about what energy 'really is' or in what form it is.

energy conservation In technology, the avoidance of waste in **energy** transfer. Energy can be wasted by being transferred to places where it is of no practical use, such as when a heated room loses energy by conduction through ceiling and walls.

energy, law of conservation of or **law of constant energy** The law that states that in any closed system the total **energy** remains constant. Another statement of the law is that energy can be neither created nor destroyed. It is one of the most fundamental laws of physics and is also known as the First Law of **Thermodynamics**. The law has been extended to include the idea that mass and energy are equivalent (*see* **relativity**) so that a change in mass Δm also means a change in energy ΔE, related by the formula $\Delta E = c^2 \Delta m$.

engine A system in which the **energy** of a fuel–oxygen system is transferred to mechanical energy. The fuel burns in oxygen to produce a large volume of hot gas. As this expands, it exerts a force and does work.

Steam engines are *external combustion engines* – the fuel burns outside the place where the energy is used. Steam *turbines* have fans driven by the hot gas.

In *internal combustion engines*, the fuel burns where the force is exerted – in the cylinder or turbine chamber. All internal combustion engines act in four stages. The diagrams opposite show these for the four-stroke petrol engine that drives most cars:

(a) *intake*: a mixture of air and petrol vapour enters the chamber through the inlet valve as the piston moves down;

(b) *compression*: the piston rises and compresses the mixture (both valves closed);

(c) *power*: a spark in the plug makes the fuel burn in the air; the hot gases that result drive the piston down;

(d) *exhaust*: the piston rises; it forces the waste gases out through the exhaust valve.

Diesels, jets and rockets are also internal combustion engines.

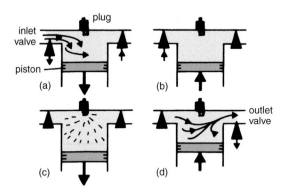

engine *The four stages of internal combusion: (a) intake; (b) compression; (c) power; (d) exhaust.*

engineering The field in which the concepts of pure science are applied to real-life problems. The branches of engineering that depend mainly on physics include:

(a) *aeronautical engineering*: the technology of aircraft and spacecraft;

(b) *civil engineering*: the design of bridges, roads, large buildings and other structures for public use;

(c) *electrical* and *electronic engineering*: dealing with devices using large or small electric currents;

(d) *marine engineering*: the design of ships and how they behave;

(e) *mechanical engineering*: the use of machines of all kinds;

(f) *nuclear engineering*: handling nuclear power systems.

environment The external surroundings, local or wider, in which people or other life-forms exist. Human activities can have a serious effect on the environment on a large scale, in a short time, and in ways which are hard to reverse. In particular, industry can have a very destructive effect on air, water, soil and rock (*see* **pollution**); there is increasing public concern about this.

equations of motion Equations that relate the following five quantities for an object moving in a straight line during a given period: its uniform acceleration, a, its displacement, s, the time taken, t, the velocities, v_1 and v_2, at the start and end of the period.

Each of the five equations relates four of these measures. Thus, given any three, it is always possible to find either of the other two.

$$s = \tfrac{1}{2}(v_1 + v_2)t \qquad \text{(omits } a\text{)}$$
$$s = v_1 t + \tfrac{1}{2}at^2 \qquad \text{(omits } v_2\text{)}$$
$$s = v_2 t - \tfrac{1}{2}at^2 \qquad \text{(omits } v_1\text{)}$$
$$v_2 = v_1 + at \quad \text{(omits } s\text{)}$$
$$v_2{}^2 = v_1{}^2 + 2as \qquad \text{(omits } t\text{)}$$

These equations come from the speed/time graph that describes this motion (*see diagram*).

The first three equations all depend on the fact that the distance travelled, s, is given by the 'area under the graph' (shaded).

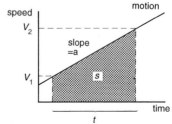

equations of motion *The speed/time graph from which the equations are drawn.*

equilibrium The state of a system in which either no changes are taking place (*static equilibrium*) or in which any that do occur are counteracted by equal and opposite ones (*dynamic equilibrium*). An object is in static equilibrium when all forces acting on it balance out so that (a) there is no resultant force, and (b) there is no resultant **moment** or turning effect. The object does not gain or lose momentum.

Examples of static equilibrium are shown in diagrams (a) to (c).

In (a), the outside forces act along the same line. The sum to one side equals that to the other. $F_1 = F_2 + F_3$.

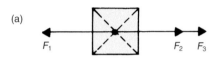

In (b), the outside forces are parallel but do not act on the same line. Again the sum to one side equals that to the other. $F_1 = F_2 + F_3$. Also the torques or moments balance so that the object does not rotate. $F_2 s_2 = F_3 s_3$.

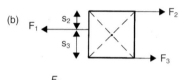

In (c), the outside forces act through one point and the **triangle of vectors** applies. Any one force is equal and opposite to the resultant of the other two. *See also* **stability**.

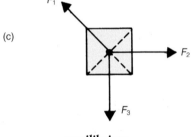

The surface of a liquid is in dynamic equilibrium when, for example, as many molecules leave the surface (evaporate) as enter the surface (condense) in a given time.

equilibrium

An object is in *thermal equilibrium* when it is neither warmer nor cooler than its surroundings. No (net) heat enters or leaves it. It is then at the same temperature as its surroundings.

erosion   The removal, by water or wind, of **rock** fragments from the place where they were broken off from an outcrop (*see* **weathering**). The fragments are crushed or worn down during transportation, until they become sand or soil. Erosion is a slow process on the whole, but greatly affects the landscape over centuries.

error   The difference between a measured value and the true value for a quantity. The 'true value' may not actually be measurable, and so is often assumed to be a mean or average of the measured values. Some errors are random, and are due to the experimental or inherent difficulty in taking accurate measurements. Others may be system errors, due to the instrument being biased in some way so that it reads consistently too high or too low.

escape speed The speed of an object leaving, for example, the surface of a planet, such that it just escapes the gravitational pull. The value for the Earth is about 11 km/s.

evaporation The escape of particles from a solid or liquid surface to form **vapour**. This can occur only if the particles have high energy, are at the surface and are moving the right way. Therefore, during evaporation there is energy transfer from the substance: its temperature tends to fall. It also follows that the evaporation rate rises with temperature rise. *See also* **boiling temperature**.

EXOR gate *See* **gate**.

expansion The increase in size of a quantity of matter. *Thermal expansion* is an increase in size owing to temperature change. Temperature rise means a rise in the particles' mean energy (*see* **kinetic model**): they move more quickly and try to take up more space. The effect is greatest with gases (*see* **gas laws**) – all gases expand by roughly the same amount when going through the same temperature change. Solids and liquids expand much less, but vary greatly in *expansivity*. The expansivity of a substance is the increase in size of a unit sample for unit temperature change. ('Size' refers to length or area for solids and to volume for solids, liquids and gases.)

$$\text{linear expansivity} = \frac{\text{increase in length } (\Delta l)}{\text{original length } (l) \times \text{temp rise } (\Delta T)}$$
$$= \frac{\Delta l}{l\Delta T}$$

Thermal expansion has many uses, as in **thermometers** and **thermostats**. However, it may also be a problem, in structures such as roads and bridges.

experiment A practical test of an idea, designed to answer a question about the subject under investigation. Experiments are an important part of the process by which science develops as part of human understanding. Indeed a subject is not a science if its study cannot involve properly designed experiments.

The design of experiments is crucial. When scientists come up with a new **theory**, or feel unhappy about one that exists already, they propose questions that experiments can answer. The hope is that the answer will either confirm or disprove the theory. Each experiment must then explore how variable measures relate to each other. *See also* **control**, **dependent variable**, **analysis**.

eye A sense organ that reacts to visible **light**. The eyeball fits in a socket in the skull, and a number of muscles can move it. The eyelids protect it; tears from a gland keep the exposed front clean and moist.

As seen in the diagram, light passes through the *cornea*, (a), into a space filled with a liquid. The *lens*, (c), hangs in a ring muscle, the *ciliary*, (d). It changes shape to focus light from objects at different distances (*accommodation*). The coloured *iris*, (b), is also a ring muscle; it opens or closes to change the amount of light passing through the central hole, the pupil.

The space behind the lens is full of jelly; this keeps the eyeball in shape.

Light reaching the layer of nerve cells forming the *retina*, (e), is absorbed. Its energy is transferred to signals which pass to the brain along the optic nerve, (f). *See also* **astigmatism, colour vision, optical instruments, vision defects**.

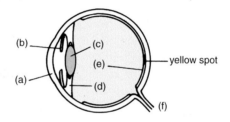

eye *The components of an eye: (a) cornea; (b) iris; (c) lens; (d) ciliary muscle; (e) retina; (f) optic nerve.*

F

farad (F) The unit of capacitance. A conductor has a capacitance (C) of 1 farad if a charge (Q) of 1 **coulomb** changes the conductor's potential (V) by 1 volt.

$$C = \frac{Q}{V}$$

Named after the English physicist Michael Faraday, (1791–1867).

fax A cheap and quick communications system that allows copies of documents (for example, letters and drawings) to be sent through the phone system using special transmitters and receivers (*fax machines*).

feedback *See* **control, oscillator.**

fibre optics The communications technology in which information is carried through thin glass fibres by light or infrared waves. The information is coded as **digital** on–off signals and may be audio signals, TV pictures or computer data. The signals are kept inside the fibre by **total internal reflection**. The combination of digitization and the high frequency of the carrier wave allows much more information to be carried more reliably by a fibre-optic system than by the copper cable system with analogue coding that it is now generally replacing in telecommunications. The diagram shows the path of the signal in a graded-index optical fibre.

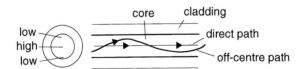

fibre optics *In a graded-index optical fibre the refractive index varies continuously. This bends light and keeps it inside the core.*

field A region of space in which **forces** appear. Thus a magnetic field is a region in which there are magnetic forces. As force is a **vector**, a field is a vector field – it has size (strength) and direction at each point. Lines of force (*field lines*) are often used to describe fields. By their direction and closeness they show the field direction and strength at each point. *See also* **electric field.**

field emission *See* **emission, electron.**

filament A thin, high-resistance wire such as the one in an electric light bulb. It is usually made of tungsten and becomes white-hot when a current passes through it. In **cathode-ray tubes** a filament heats the **cathode** and sometimes acts as the cathode. Electrons escape and are attracted to the **anode**.

filter **1.** (in optics) A piece of material allowing only one band of wavelengths to pass. The others are absorbed. Thus a green filter passes only green. *See also* **absorption, colour**.
   2. An electronic circuit which passes some frequencies of **alternating current** (ac) better than others.

fission The splitting of an unstable nucleus into two roughly equal-sized parts which are also nuclei and are called the *fission products*. Some large nuclei undergo fission spontaneously, for example, uranium-235 and uranium-238. These nuclei, and others, may also be split as a result of absorbing a neutron (*artificial fission*). Fission may also release other particles, such as neutrons and high-energy photons (gamma radiation), and give kinetic energy to the fission products. In general, when large nuclei split, the masses of the product nuclei are less than the mass of the original nucleus. The missing mass appears as energy, and fission is the source of the energy used to generate electricity in nuclear power stations. The diagram shows what happens when a uranium-235 nucleus splits. *See also* **chain reaction, atomic bomb**.

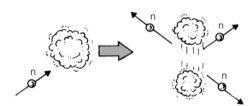

fission *The fission of a uranium-235 nucleus.*
$$^{235}_{92}U + ^{1}_{0}n \rightarrow ^{236}_{92}U \rightarrow \textit{fission products} + 3^{1}_{0}n + \textit{energy.}$$

fixed points or fixed temperatures The standard temperatures that can be accurately reproduced and so used as the basis of a temperature scale. This means that they can be used to help **calibrate** thermometers. The best-known fixed points are the *ice point*, the temperature of pure melting ice (0 °C, 273.15 K) and the *steam point*, the temperature of pure boiling water at standard atmospheric pressure (100°C, 373.15K).

flip-flop Another name for a **bistable** multi-vibrator switching circuit.

flotation The ability of an object to float (*see* **buoyancy**). A floating object displaces its own weight of fluid. This follows from **Archimedes' principle**. *See also* **upthrust**.

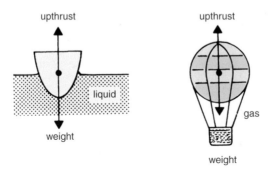

flotation

fluid A substance, such as a liquid or a gas, that can *flow* because its particles are not fixed in position like those of a solid.

fluid friction *See* **viscosity**.

fluorescence The emission of light or other **radiation** from atoms bombarded by radiation. It is a fairly common result of **absorption** of radiation. When a certain substance absorbs a certain radiation, the energy appears as radiation of a lower frequency. Fluorescent paints and fluorescent tubes contain chemicals which absorb **ultraviolet** light and release the energy as visible light. Some washing powders have such a chemical added so that clothes appear extra white in daylight. See **discharge**. Fluorescent lamps turn more of the electrical energy to light compared with filament lamps, and so are about three times more efficient (*see* **discharge** and **efficiency**).

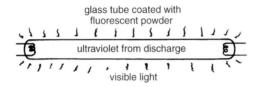

fluorescence *A fluorescent lamp.*

focal distance or focal length (f) The distance between the geometric centre (the **pole**) of a **lens** or **mirror** and the *focal point*; the shorter the focal distance, the greater the **power** of the lens to bend parallel rays of light. The focal point of a lens or mirror is the point toward which, or from which, light, travelling close to the axis and parallel to it, moves after **refraction** or **reflection**.

f appears in the *lens/mirror equation*:

$$1/f = 1/u + 1/v$$

Here u is the distance between the pole and object, and v is that between the pole and **image**. In this equation, a real focal point is denoted by f^+, an imaginary (*virtual*) focal point by f^-.

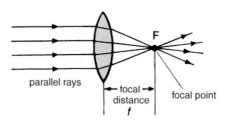

parallel rays

focal distance f

focal point

F

focal distance

focal point *See* **focal distance**.

force (F) Anything that tends to change an object's **momentum**, i.e. to make it move more or less quickly and/or to make its direction change. The unit is the newton, N: the force that will accelerate one kilogram by one metre per second per second.

The forces we meet in everyday life are based upon two of four fundamental forces of nature:

(a) *gravitational*: a force of attraction only, acting between masses;
(b) *electromagnetic*: the forces of attraction and repulsion between electric charges (*electrostatic*) and the forces of attraction and repulsion due to moving charges (*magnetic*).

These forces 'act at a distance', i.e. through space. Many forces act when there is, apparently, contact between the objects experiencing force. This occurs in friction, push and pull (compressive and tensile) forces, and twisting. In fact, however, contact does not occur: the forces are electromagnetic and arise from the electric charges making up **atoms**.

The other two types of force are the *strong nuclear force* and the *weak nuclear force*, which are very short-range forces that act on **fundamental particles** (for example, in the nuclei of atoms).

force meter or **newton meter** A device used to measure **force**. The many kinds of spring **balance** are force meters – the unknown force is balanced against the stress in a spring. The spring's change of length per unit applied force is known from tests.

force ratio (for a **machine**) The force overcome (output force, or load) divided by the force applied (input, or effort). Force ratio may be called **mechanical advantage**. It has no symbol or unit.

fossil fuel Any fuel produced by organic matter, formed from dead organisms that, although buried, have for some reason failed to decay (oxidize) completely. Fossil fuels take a long time to form, and so, on the time scale on which they are being used up, they are not considered renewable. Coal, oil and natural gas, and their products (for example, petrol and town gas) are fossil fuels, as is peat (plant remains in the process of being turned into coal). The world's supply of these fuels will not last forever, so there is worldwide concern to find alternative sources of energy before they are used up. Another problem is that they burn to produce gases that increase the **greenhouse effect**.

free fall The downward motion of an object in which the only outside force acting on it is **gravity**. It then appears **weightless**. The Earth is in free fall round the Sun; so is a massive rock dropping down a cliff. In cases such as these, air **friction** can often be ignored. An object in free fall moves with constant acceleration, g. Near the Earth, g is about 10 m/s². *See also* **weight**.

frequency (f) The number of cycles of a periodic motion in unit time; for instance, the number of waves passing a point in a second, or the number of swings of a pendulum in a minute. The unit is the *hertz* (Hz), one cycle per second.

The frequency, f, **wavelength**, λ, and speed, v, of a wave relate to each other in the *wave equation*:

$$v = f\lambda$$

See also **period**.

friction A force tending to prevent the motion of bodies whose surfaces are in contact. Sometimes friction is useful and works in our favour. For instance, friction tends to prevent feet from slipping while walking or a car from skidding when turning.

Friction between wheels and road or track is essential for road vehicles and trains, otherwise the wheels would slip and there would be no propulsion. Once the vehicle is moving, air friction (*drag*) is the main frictional force opposing the driving forces and energy is wasted in doing work against the drag force. At maximum speed the driving force equals the frictional force (*see* **terminal speed**). Cars, trains and aircraft are designed to reduce drag, so that as little energy as possible is wasted. Friction in internal moving parts also causes wear.

If a block placed on a surface is pulled with a gently increasing force P as shown in the diagram, at first the block does not move: the friction force F automatically adjusts itself to equal P. There is, however, a value of P at which the block almost starts to move. F is then the *static friction force* (or *starting friction force*). Once the block starts to move, the value of P that keeps it moving is less than before. F is now the *dynamic friction force* (or *sliding friction force*). The force of friction between two solid surfaces does not depend on the area of the surfaces 'in contact'. This is because at the microscopic level the surfaces are not smooth, so they do not touch over the whole of their contact area. The frictional force does, however, depend on the force with which the two objects are pressed together (the *normal reaction*, R):

$$\text{friction force } P = \mu R$$

where μ is a constant, the *coefficient of friction*.

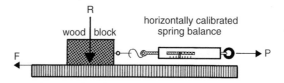

friction *Static and dynamic friction.*

fulcrum *See* **lever.**

fundamental particle One of the large number of subatomic particles that make up the matter present in the universe. Ordinary matter (such as the atoms of chemical elements) contains **electrons**, **protons** and **neutrons**. Protons and neutrons are in turn made of smaller particles called *quarks*, held together by *gluons*. There are also many other particles which are produced in stars and energetic galaxies, and, on Earth, in high-energy particle **accelerators**. The most common of these is the *neutrino*, which has a very small mass and no charge. It is now usual to include massless particles (for example **photons**) in this category. *See also* **particle physics.**

fuse **1.** To melt.

2. A safety device in electric circuits: a length of copper or an alloy melts when the current becomes too large because of a fault. This protects the cable from damage and also cuts off the device from the live wire in the supply. Fuses depend upon the heating effect of an electric current.

Household fuses come in various 'ratings': 30A, 15A, 5A and 3A, with smaller 1A fuses for some electronic devices. The lowest rating of fuse required by the device should be chosen so that quite a small increase in current will 'blow' the fuse. This can be calculated using the power formula in the form $I = P/240$, where I is the current taken by a device of power rating P. For example, if I is calculated to be 2A then a 3A fuse should be chosen, not a higher-rated fuse. Fuses are found in the junction box which contains the house electricity meter, and many devices are also fitted with mains plugs which contain a fuse of the correct rating. Circuits must be switched off and/or devices disconnected from the mains before fuses are replaced.

See also **mains plug, circuit breaker, conduction, electrical.**

fusion 1. Melting (see **state change**).

2. The joining of two or more light atomic nuclei to make a more massive one. There is a **mass–energy** transfer in this case. Fusion is used in the 'hydrogen' bomb; effective ways to control it and produce inexpensive clean power have not yet been found.

Here is a simple fusion reaction, between two **deuterium** (heavy hydrogen) nuclei:

$$^2_1H + ^2_1H \rightarrow ^3_2He + ^1_0n + \text{energy}$$

The two deuterium nuclei have more mass than the helium nucleus and the neutron. The missing mass appears as kinetic energy of the product particles.

The energy of **stars** comes from fusion.

G

G The symbol for the universal gravitational constant in Newton's law of **gravity**. *See Appendix B.*

gain The ratio of the output signal strength of a **circuit** or system to the input, a measure of amplification. In practice, the output and input power, voltage or current are compared. The normal unit of gain is the **decibel** (dB) – a gain of 3dB means that the output is twice the input; with a gain of –6 dB the output is a quarter of the input. *See **amplifier**, **bel**.*

galaxy A group of millions or billions of **stars** held together by their own gravitational attraction to each other. Galaxies are very large; an average galaxy contains 100 billion (10^{11}) stars, and is 100,000 light-years in diameter. Much – perhaps most – of the matter in a galaxy is invisible. Much unseen matter consists of thin clouds of gas (hydrogen and helium), but there may be much larger quantities of an unknown form of matter. Galaxies also contain dust and small grains of rock, ice and frozen gases, which shut out the light emitted by stars. There are two main types of galaxy, defined by their shape: *elliptical* and *spiral* galaxies. Elliptical galaxies seem to be older – they contain a larger fraction of older stars well on in their life-cycle. Spiral galaxies tend to contain younger stars, and often very young stars are observed in the spiral arms, where it is believed stars are currently being formed from clouds of gas and dust.

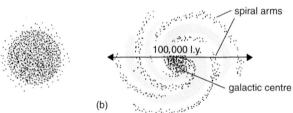

galaxy *(a) An elliptical galaxy; (b) a typical spiral galaxy.*

The Sun is in the spiral arm of a galaxy called the *Milky Way*. The Milky Way is 100,000 light-years in diameter. The Sun is about 25,000 light-years from the centre, which we cannot see because it is hidden by clouds of gas and dust. The Milky Way is easy to see as a broad band of light crossing the night sky from the constellation of Canis Major to Sagittarius, via Perseus, Cassiopeia and Cygnus. Binoculars will show that the light comes from a very large number of stars, seemingly close together in the sky but in reality separated by immense distances. Most galaxies occur in clusters of just a few to several thousands of galaxies. The Milky Way is part of a small cluster of about 30 galaxies.

Galileo Galilei (1564–1642) An Italian scientist who was one of the first to break away from the ideas of **Aristotle** and others handed down from ancient times. He was the first person to use a telescope for research; his new theories of **dynamics** and the structure of the universe led to conflict with the Church.

galvanometer A sensitive instrument for detecting or measuring small electric currents. A milliammeter is an example of a galvanometer (*see* **ammeter**). The galvanometer is named after the Italian scientist Luigi Galvani (1737–98).

gamma radiation (γ) The electromagnetic radiation emitted as a result of radioactive decay. Gamma rays have high energy and are emitted as separate pulses or particles of radiation – **photons**. They are the most penetrating and the least ionizing of the three radiations emitted in **radioactivity** (alpha, beta and gamma). The emission of a gamma ray does not change the nature of the emitting nucleus; it is due to a change in its internal energy state. *See also* **atomic radiations**.

gas One of the three states of matter. A gas has no fixed shape but takes up the shape of its container and fills it completely. The particles of a gas (usually **molecules**) move randomly at high speed and cause pressure on the walls of their container as a result of continuous bombardment. The particles are further apart than when in the **solid** or **liquid** state, so a gas is the least dense state of matter – about a thousand times less dense than the solid or liquid state of the same substance.

Gases result from the **evaporation** of solids and liquids – the sublimation of solids and the boiling of liquids. A gas differs from a **vapour** in that pressure alone cannot convert it to liquid. *See also* **critical temperature, kinetic theory, gas laws**.

gas laws The laws that relate the **pressure**, p, the **volume**, V, and the absolute **temperature**, T, of perfect or 'ideal' gas samples. There are three laws; each concerns a fixed mass of gas:
(a) **Boyle's law**: For a fixed mass of gas at constant temperature, the product of pressure and volume is constant.
(b) **Charles' law**: For a fixed mass of gas at constant pressure, the volume is proportional to the absolute temperature.

(c) *Pressure law*: For a fixed mass of gas at constant volume, the pressure is proportional to the absolute temperature.

If for some reason, a gas sample moves from one state, p_1, V_1, T_1, to a second, p_2, V_2, T_2, the laws can be written thus:

Boyle	$p_1V_1 = p_2V_2$	(T constant)
Charles	$\dfrac{V_1}{T_1} = \dfrac{V_2}{T_2}$	(p constant)
Pressure	$\dfrac{p_1}{T_1} = \dfrac{p_2}{T_2}$	(V constant)

They combine into the *ideal gas equation*; this also applies to constant mass amples:

$$\frac{p_1V_1}{T_1} = \frac{p_2V_2}{T_2}$$

All these statements apply only to ideal gases. In practice, real gases follow them fairly closely at normal temperatures and pressures.

gate A microelectronic circuit or device used in control and logic applications. Gates are made from combinations of transistors and resistors behaving as switches that can be either on or off – called low or high in control systems and logic 0 or 1 in logic and computer applications. Gates have inputs and outputs; the output is decided by the input state. The main gates in use are as follows:

AND *gate*: a two-input gate whose single output is high (1) only when both inputs are high. See diagram for a simple model of an AND gate.

The AND gate.

In the circuit shown, the LED is 'on' only when both 'A' AND 'B' are 'on'. Hence the name AND gate.

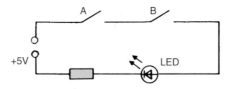

gate *A simplified electrical circuit using an AND gate.*

OR *gate*: a two-input gate whose single output is high (1) when either or both inputs are high.

The OR gate.

	AND			NAND			NOT	
A	B	O	A	B	O	A	B	
0	0	0	0	0	1	1	0	
1	0	0	1	0	1	0	1	
0	1	0	0	1	1			
1	1	1	1	1	0			

	OR			NOR			EXOR	
A	B	O	A	B	O	A	B	O
0	0	0	0	0	1	0	0	0
1	0	1	1	0	0	1	0	1
0	1	1	0	1	0	0	1	1
1	1	1	1	1	0	1	1	0

gate *Truth tables.*

NOT *gate*: a single-input gate whose single output is high when the input is low, and low when the input is high.

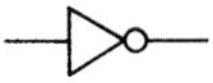

The NOT gate.

NAND *gates* (not AND) and NOR *gates* (not OR) act as the opposite (inverse) of AND and OR gates.

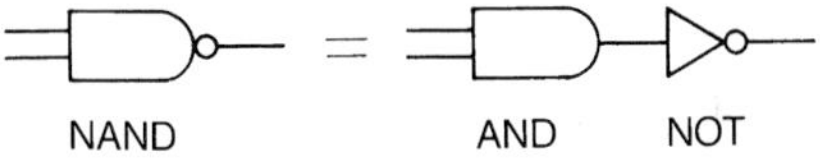

NAND = AND NOT

The NAND gate.

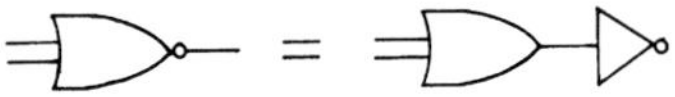

The NOR gate.

EXOR *gates* (exclusive OR) produce a high output if, and only if, *one* input is high.

The behaviour of gates is best described by *truth tables*, which list all the possible combinations of inputs and outputs. The truth tables and symbols for the gates listed here are given opposite; A and B are inputs, O is output.

gear A toothed wheel whose function is to transfer **moment** (*torque*) or turning motion from one shaft to a second. A gear system is a type of **machine**. The torque ratio equals the *gear ratio* – the ratio of the numbers of gear teeth on the input and output wheels.

The diagram shows two of the simplest systems. The *spur gear*, (a), is used to reverse the torque or turning speed between two parallel shafts. In (b) the output shaft turns the same way as the input shaft.

In practice, to reduce slip, the teeth have a much more complex shape than that shown.

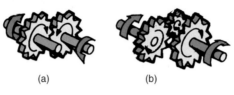

(a)   (b)

gear

Geiger–Müller tube A sensitive **radiation** detector. It contains argon gas at low pressure mixed with bromine. There is a **potential difference** of 350–450 volts between the anode and cathode. *See diagram below.*

A particle of ionizing radiation produces a trail of **ions** in the gas. The electric field in the tube causes the ions to move towards the electrodes. The light negative electrons accelerate at a high rate; as they collide with gas particles, producing more ions, and so on. Thus each packet of input radiation causes a shower of charge in the tube and a pulse of current in the output circuit. The pulses may drive a speaker, **counter** or **scaler**.

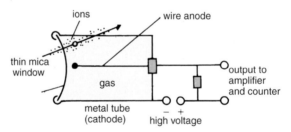

Geiger-Müller tube

generalization Making a theory (hypothesis) that what you observe in one or a small number of cases applies more widely. For instance, after studying how one acid reacts with a metal, you may suggest that all acids react with the metal that way, or even that all acids react that way with all metals. Either of those generalizations is a valid step in science, providing someone then designs careful **experiments** to test the theory.

generator A machine that produces an **electromotive force** as a result of a coil spinning in a magnetic field, or by a magnetic field moving near a coil. In the large generators used in power stations, the coils are spun at high speed by turbine wheels. The force to move the turbines may come from superheated steam (in coal, oil or nuclear power stations) or from water under pressure (*see* **hydroelectricity**). The emf is produced by **electromagnetic induction** which causes an alternating voltage between the ends of the coil. The diagram shows a simple generator – in practice, several coils are used.

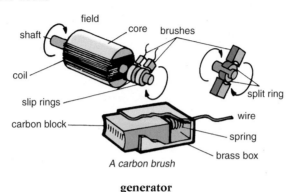

A carbon brush

generator

In an alternating-current generator, the current is taken from a generator via *slip rings*, connected to the spinning shaft, and carbon *brushes*, connected to the external circuit. The brushes are soft and press against the copper slip rings to make good contact. A single *split ring* will provide unsmoothed direct current as shown in the diagram.

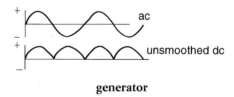

generator

geological time   A time-scale suitable for describing the history of the Earth since it formed thousands of millions of years ago. The time is divided into a number of 'ages' and 'periods', based mainly on the life then existing on Earth. Life appeared at the end of the Precambrian period, about a thousand million years ago. We are now in what is called the Holocene part of the Quaternary period. *See Appendix F.*

geostationary or geosynchronous satellite   An Earth **satellite** that orbits the Earth at just the right height for its orbital period to equal the daily rotation period of the Earth. If its orbit lies over the equator, the satellite stays fixed over a given point on the Earth at a height of 35,900 km.

If the orbit is tilted, the satellite seems to move north and south of the equator throughout the day, as viewed from the ground. Geostationary satellites are generally used as re-transmitters of TV and telephone signals, and their development has made 'satellite TV' possible. Geostationary orbits are also used for weather satellites.

geothermal energy The heat energy in the rocks below the Earth's surface. Sometimes it escapes in a volcano. Sending water down boreholes can release this energy as steam to run a power station and provide electricity.

global warming A predicted result of the **greenhouse effect**, in which increasing quantities of industrially produced carbon dioxide (and other pollutant gases) in the atmosphere act as a blanket to radiation leaving the Earth. This is predicted to produce a gradual increase in the average temperature of the Earth, leading to expansion of the oceans, melting of ice caps and glaciers, and the alteration of climatic and weather patterns. These effects may produce flooding of low-lying land or change the kind of crops that can be grown in certain regions. Evidence that the climate really is warming is now strong, but whether it is due to human activities is much more controversial.

globular cluster A spherical or near-spherical group of stars held close together by their mutual gravitational attraction. A globular cluster may contain from many thousands to several millions of stars. Most of the globular clusters so far detected are in a large sphere surrounding our **galaxy** (the Milky Way) – there are about a hundred of them and several can be seen with a small telescope or a pair of binoculars.

gradient Another word for slope. *See* **graph**.

graph A method of showing **data** pictorially. Patterns can be seen more clearly in a graph than when the data are in table form.
    Graphs show how two variable quantities (more than two in some cases) relate to each other. Thus graph (a) shows that y is **proportional** to x. In (b) y is inversely proportional to x. In physics it is best to obtain graphs like those in (a) or in (c) – straight lines, sloping up or down, perhaps passing through the origin, O (the point $x = 0$, $y = 0$). Such graphs show most clearly how x and y relate.
    It is often useful to find the *slope* and intercept of a graph. The slope, or *gradient*, m, is the ratio $\Delta y / \Delta x$ (as in the diagrams opposite). The *intercept*, c, is the value of y at which the graph crosses the y–axis; this is the value of

y when *x* is zero. The values of *m* and *c* in the case of a straight line graph give the equation of the line: $y = mx + c$.

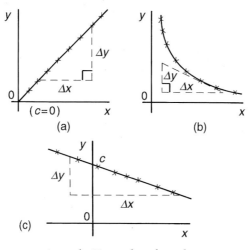

graph *Examples of graphs.*

There are other types of graph, of more value in different situations. The **histogram** (bar chart) and pie graph are the most important of these other graphs.

graphite A form of carbon that conducts electricity well. It is a soft crystal and so is used as a lubricant. It is also used to make pencil 'leads' and as a *moderator* to slow down neutrons in a **nuclear power** reactor.

gravity One of the basic types of **force**. Gravity is an attraction between masses. The force *F* of attraction is given by the formula:

$$F = \frac{G\,Mm}{r^2}$$

M and *m* are the masses and *r* the distance between them. *G* is the gravitational constant. Physicists are building sensitive detectors with which they hope to observe gravitational waves; the existence of such waves would imply the existence of corresponding particles, which have been christened gravitons. *See also* **centre of gravity**.

greenhouse effect 1. The ability of glass or plastic to trap energy when radiation shines on them and so to warm up the space behind them. This contributes to the warming of the air inside a greenhouse.

2. The ability of the atmosphere of the Earth or of another planet to trap radiation energy and so help raise the planet's temperature. 'Greenhouse materials' are transparent to high-frequency radiation but opaque at low frequencies. This means that in both cases high-frequency (short-wavelength) electromagnetic radiation from the Sun can enter and warm up the ground and other objects. These will also emit radiation but with lower frequency (longer wavelength), which is absorbed by the glass or plastic or the atmospheric gases, which thus act as a kind of blanket to reflect thermal radiation back to the surface of the Earth. The most effective greenhouse gas found in the atmosphere is carbon dioxide. Carbon dioxide is produced by the respiration of living things but absorbed by plants during photosynthesis.

These two processes of respiration and photosynthesis kept a fairly steady concentration of carbon dioxide in the atmosphere for millions of years. But burning organic matter (for example, wood, oil) also produces carbon dioxide and in the last two hundred years or so the increasing use of fossil fuels has increased the concentration of carbon dioxide in the atmosphere. It is feared that this will result in a slow warming of the atmosphere – **global warming** – with a rise in sea level. The melting of floating ice in the polar icecaps would not alter sea levels, but if ice sheets now on land were to melt partially and slip into the oceans, there could be additional rises. If this happens, ports and many low-lying areas of the world may become permanently flooded.

Methane is also a greenhouse gas and is produced in large quantities by cattle. As the world's population increases, more and more cattle are farmed and this also contributes to the increase in the greenhouse effect.

ground station A place where communications and data signals from weather, communications and monitoring satellites are received.

H

half-life (T½) The time in which a measure whose **decay** is exponential falls to half its value. The unit is the second, *s*. If the half-life of a radioactive substance is 5 seconds it will decay to half its activity in 5 seconds, one-quarter in 10 seconds, one-eighth in 15 seconds, etc.

In the graph, *y* may be, for example, the number of radioactive nuclei in a sample, the voltage between the ends of an **inductor**, or the charging current of a **capacitor** after switch-on.

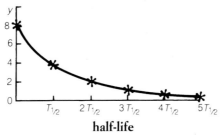

half-life

hardness A measure of how well a surface resists scratching. There is a 10-point scale for this, the *Mohs scale*, based on comparisons with the known hardness of various minerals. On the scale, talc (value 1) is the softest, diamond (10) is the hardest.

hearing The use of a specialized sense organ (the **ear** in higher animals) to:
(a) collect sound waves in a certain frequency range;
(b) convert them to signals to pass to the brain;
(c) use these to obtain information from outside.
In this way, the healthy human ear can handle sound waves that range from extremely soft to very loud, and in frequency from about 20 Hz to over 20,000 Hz. *Hearing defects* include partial deafness (where one cannot detect weak sounds, or where one can hear only a small range of frequencies), to total deafness. Some ear problems produce a permanent ringing sensation; others affect the sense of balance (for the ear is also responsible for this).

heat 1. The everyday name for the internal energy of a substance that determines its **temperature**. This internal energy is due to the motion of its particles. In a solid, the particles (atoms, ions) vibrate about fixed points, so that the energy is a mixture of **kinetic** and **potential energy** (the potential energy arises from the forces that hold the particles together).

Particles in liquids and gases can move more freely; in a gas most of the internal energy is kinetic energy (*see* **kinetic model**). Internal energy can be transferred from a body at a high temperature to one at a lower temperature by the *thermal processes* of **conduction**, **convection** and **radiation**. Note that 'heat' is a form of energy but temperature is an energy *level* or 'degree of hotness'. As a general rule for solids and liquids, increase in internal energy (*E*) is proportional to increase in temperature (*T*):

$$\Delta E = ms\Delta T$$

where *m* is the mass of the substance and *s* is a constant for that substance called its **specific thermal capacity**.

2. To cause an *energy transfer* from one body to another by a difference in temperature between them that increases the internal energy of the heated body (by increasing its temperature or changing its state in some other way). For example, water can be heated by thermal conduction through the metal of a saucepan, and this raises its internal energy and its temperature.

heat capacity *See* **thermal capacity.**

heat engine A device that uses the processes of thermal transfer of energy from a hot object to a cooler one to do work. The most common type of heat engine is the internal combustion engine as used in cars.

heavy water *See* **deuterium.**

hertz (Hz) The unit of frequency, one cycle per second, named after Heinrich Hertz (1857–94), a German physicist who discovered radio waves in 1886.

Herzsprung–Russell (H–R) diagram A graph in which certain characteristics of stars are plotted. The plot may be of *luminosity* (energy output via radiation) against *surface temperature*, or of the related quantities of *absolute magnitude* against *stellar class*. Most of the stars seen in the night sky fall in a broad band called the *main sequence* (*see diagram opposite*). The main sequence crosses from hot bright stars (usually large) at the top left to smaller, cooler stars at bottom right. It is conventional to plot temperatures decreasing from left to right.

The main sequence consists of stars that get their energy from converting hydrogen into helium by **fusion** reactions. When a star is formed, it appears on the H–R diagram to the right of the main sequence, because fusion reactions have not yet begun. When hydrogen- 'burning' begins, the star enters the main sequence at a particular place and stays there for most of its life. Some stars expand very suddenly towards the end of their lives and become much larger and cooler: they move to the upper right as *red giant* stars. Other stars with less mass do not expand but shrink and become hotter, appearing as less bright stars (they are very hot but too small to emit much radiation), so that they move to the bottom of the diagram. They are called *white dwarfs*. Red giants may also evolve to become white dwarfs.

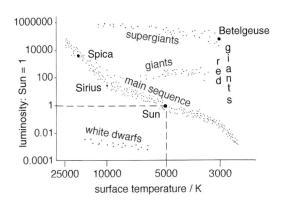

Herzsprung-Russell diagram

histogram A chart used mainly to show the numbers of items in each group of some numerical data. It is in fact a frequency **graph** based on continuous data. The histogram given plots the graph of the number of a company's outgoing telephone calls against duration.

A *bar chart* is used when one set of data is not numerical, for example, the number (numerical) of organisms of different species (non-numerical) found in a pond. It is sometimes useful to form a line graph from a bar chart, which can be done by joining the mid-points of the tops of the bars.

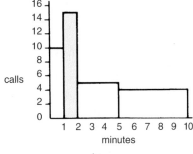

histogram

holes Sites of missing **electrons** in the structure of a solid **semiconductor**. When there is a **potential difference** between the ends of a sample, electrons can move to fill the holes and in effect the holes move the other way to form an electric current.

holography The technology of recording a three-dimensional image of an object onto film using **laser** light. When developed, the film shows different views of the object from different angles. Modern holograms can be viewed in normal light. They can also show animated 3-D images of large objects in full colour.

Hooke's law A law that states that the extension or compression of an object is proportional to the force applied.

This law applies only approximately, and only to elastic materials, i.e. those that return to their original size when the applied force is removed. Some elastic materials (like rubber) obey Hooke's law only up to a certain extension, after which the increase in length stops being proportional to applied force; *see graph* (a). Coil springs are designed to obey Hooke's law, and so may be used as force measurers (newton meters) and spring balances. For such objects the equation is:

$$\text{extension} = \text{constant} \times \text{force}$$

Metals tend to obey Hooke's law for small extensions, after which they reach an *elastic limit* and begin to behave *plastically*. This means that they become permanently stretched. *See also* **elasticity**.

Hooke's law *(a) The behaviour of rubber. The rubber will return to its original length so is elastic. (b) The behaviour of a metal such as copper.*

Robert Hooke (1635–1703) was a British scientist in the group that founded the **Royal Society**.

Hubble's law A law deduced by the American astronomer Edwin Hubble from his observations of the speeds and distances of **galaxies**. He noted that the further away the galaxies were, the faster they were moving away from the Earth. This is now understood as an expansion of space on a large scale (carrying its contents with it), so that every point is gradually moving further away from every other point. Hubble's law provides a simple formula: $v = Hr$, where v is the speed at which two galaxies a distance r apart are moving away from each other. H is a constant, the *Hubble constant*. The precise value of H is not known, but it lies between 15 and 30km per second per million light-years.

hydraulic machine A machine that multiplies a force by using a liquid to transmit pressure from one place to another. Examples include a hydraulic jack, a hydraulic press and hydraulic brakes. Hydraulic systems are used widely in large machines such as bulldozers, earth movers and mechanical diggers, and in moving the control surfaces in aircraft and ships.

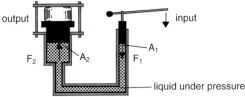

hydraulic machine *A hydraulic press.*

For example, in a hydraulic press, a small force acting on a small area (A_1) produces a large pressure in the hydraulic fluid (usually oil). The same pressure acts on a large surface (A_2), so producing a large force (thrust). By definition of pressure:

$$\frac{F_1}{A_1} = \frac{F_2}{A_2}$$

so that:

$$F_2 = F_1 \times \frac{A_2}{A_1}$$

hydroelectricity Electric power produced by the energy of water falling, perhaps from behind a dam. Such power stations can be costly to build; on the other hand, they are very cheap to run. They also allow a form of large-scale electricity storage: in a *pumped storage system*, power from the grid at night returns water behind the dam. More power is then 'on tap' during peak demand times. *See also* **electricity generation, generator.**

hydrogen bomb *See* **fusion.**

hydrometer A device used to measure a liquid's relative **density**. It can be used to check the state of car batteries and of milk and other drinks.

hypothesis An explanatory idea, based on reasoned argument and/or observational data, which makes predictions concerning the outcome of an event, experiment or test. However, it is a scientific hypothesis only if it is possible to disprove it; this means it must be possible to check it with further experiment or observation. The hypothesis becomes a theory, and then a law, if no one disproves it after some length of time (but even then it may still be disproved or extended later).

I

ice temperature *See* **fixed temperatures**.

ideal gas A 'model' gas that obeys the **gas laws** perfectly. The model is based upon the calculations of the **kinetic theory** of gases. Real gases tend to behave like ideal gases at low pressures, when the particles are far enough apart not to affect each other too much.

illumination (*I*) A measure of the energy intensity (brightness) of light reaching unit area of a surface. It depends on the brightness, distance and angle of each source. The unit is the lux (lx).

image The picture of an object made by an optical system, such as a lens, mirror, telescope, microscope, camera, etc. The image is *real* if light actually passes through it. Real images can be captured on a screen or on film in a camera. The image is *virtual* when the light, after being redirected by the optical system, appears to come from the image, but does not actually pass through it. The everyday example of a virtual image is the image in a looking-glass (*see diagram*).

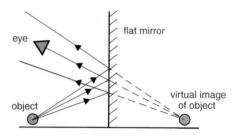

image *Virtual image.*

impulse The product of force (*F*) and the time (*t*) for which it is applied, which equals the change of momentum caused (ΔP). Impulse is a particularly useful idea when the time for which the force acts is small. The formula is $Ft = \Delta P$.

independent variable *See* **dependent variable**.

induced magnetism The magnetism produced in a material when it is placed in a magnetic field. The material becomes magnetized in the same direction as the field lines.

inductor An electric circuit component, usually a coil of wire wound around a soft iron core, in which an electromotive force (emf) is induced by a changing magnetic field. The changing field may be produced by a current in a nearby coil, or by a current in the coil itself (when the process is

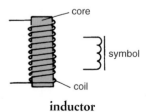

inductor

known as *self-induction*). Self-induction produces an emf that opposes the change of current in the coil, i.e. it produces a *back-emf*. *See also* **electromagnetic induction**.

inertia The property of mass that makes a body hard to accelerate or decelerate. *See also* **momentum, Newton's laws of motion**.

inference A process of reasoning in which a **conclusion** is reached, based on certain facts obtained by experiment or observation. An inference may lead to a **generalization**, the basis for further experiment and theory.

information technology (IT) The use of modern techniques to store, transfer and process **information**, as a tool for life. Computers, phones, TV and audio systems are all examples.

infrared A region in the spectrum of **electromagnetic waves**. Roughly, the wavelength range is 10^{-6} to 10^{-3} m, and the frequency range is 3×10^{11} to 3×10^{14} Hz.

All matter emits infrared **radiation** at all times – it is 'thermal radiation' (*see also* **heat**), detectable by thermometers, human skin nerve cells, thermopiles and some photographic films. Infrared radiation is now widely used in communication systems, for example in remote control units for TV sets, video recorders, etc.; carrying conversations and data along optical fibres; transferring data between computers and printers without the use of cables.

infrasound **Sound** of frequency too low to affect the human **ear**, below about 20 Hz. The vibrations may still be 'felt' by the body. *See also* **audible frequencies**.

insulation Material that cuts down the transfer of energy or charge from one place to another.

Thermal insulation reduces energy transfer by thermal processes (**conduction, convection, radiation**) and can be used to keep the contents of, for example, food containers hot or cold. Many insulators are materials with small air pockets (like wool in clothes, rockwool or glass fibre in loft insulation). The air is a poor thermal conductor, and because it is contained in small pockets, convection is inhibited. Shiny surfaces help in heat insulation because they reflect radiation and are also poor radiators. *See also* **vacuum flask**.

Sound insulation cuts down energy transfer (and so noise) by sound waves, either by reflection or, more usually, by absorption.

Electrical insulation uses non-metals, usually plastics but sometimes ceramics and glass, to insulate conductors at different voltages. These materials do not have charged particles that are free to move. They are vital for constructing circuits which do not 'short out', and of course, for safety, isolating conductors at high voltages from making contact with people. *See also* **double insulation**.

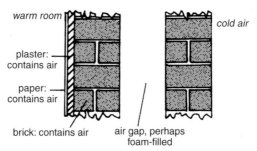

insulation *Cross-section of an external wall showing insulating features.*

integrated circuit (IC) A device that consists of a number of circuit elements formed in the surface of a chip of **semiconductor** material (silicon in most cases).

Modern ICs may contain millions of circuit elements on a chip only a few millimetres square. They are very cheap and robust, use very little power, and are fast-acting.

As a result of progress in such microelectronics, a relatively inexpensive modern pocket **computer** has the power – and none of the problems – of the early machines that cost millions of pounds and needed large rooms in which to operate.

intensity The rate of **radiation** energy transfer per unit area. The intensity of a sound wave relates to its loudness; that of a light wave gives the brightness. In each case, it is the **amplitude** of the **wave** that determines intensity. In general, intensity is proportional to the square of the amplitude. The unit of intensity is watts per square metre (W/m^2). *See also* **illumination**.

interference The effect produced when two (or more) waves meet at the same place, and interact. Waves obey the principle of *superposition*, which means that the effects of a number of waves simply add up. For example, water waves are caused by a displacement of the water surface from an average position (*see opposite*). When two equal crests meet, the result is a crest of double height (diagram (a)i). When the crest of one wave meets the trough of the other, the result is zero displacement (diagram (a)ii). The waves are unaffected by this meeting; they simply pass through each other and travel on.

The classic example of interference is the pattern produced by waves which have passed through two narrow slits. Diagram (b) shows the action of the waves, which can be demonstrated by using water waves in a **ripple tank**. The wave–no wave pattern is a result of *constructive* and *destructive interference*.

Diagram (c) shows **Young's two-slit experiment** for interference of light waves. This experiment works best when the two interfering beams have the same brightness and wavelength and there is only a small difference between the lengths of the paths travelled by the two beams. A laser produces the clearest results. *Young's fringes* are produced by constructive and destructive interference. Waves arrive at a particular place always either 'in phase' (crest meets crest, and then trough meets trough) or 'out of phase' (crest always meets trough).

Similar interference effects can be produced using microwaves and using sound waves (from two loudspeakers emitting the same frequency).

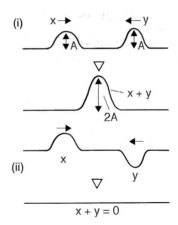

interference *(a) The principle of superposition.*

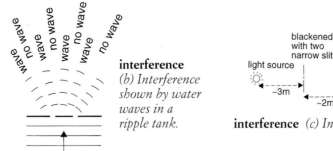

interference *(b) Interference shown by water waves in a ripple tank.*

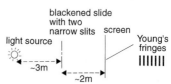

interference *(c) Interference of light.*

internal energy The sum of the energy possessed by the particles in a body due to their relative motions and potential energies. This term is preferred in advanced work to the terms **heat** or *thermal energy*. The internal energy of a body can be changed by doing work on it (for example, compressing a gas) or by heating it (i.e. by allowing energy transfer from a hotter body via thermal processes).

internal reflection *See* **total internal reflection.**

internal resistance The resistance inside a source of electricity due to its structure. A source of electricity (for example, a **generator**) provides a voltage (**electromotive force**) that drives an electric current through a circuit but also through itself. There will be a **resistance**, however small, in the source; therefore the output voltage is smaller than the true emf because of the voltage loss due to the **resistance** of the source.

inverse square law A law that holds in many areas of physics, stating that the **intensity** of an effect is reduced in inverse proportion to the square of the distance from the source. For instance, the light and heat energy from a fire falling on a square metre of surface is reduced if the distance between the surface and the fire is doubled, to $(\frac{1}{2})^2$ of its previous value, $\frac{1}{4}$. This law is true for most types of radiations and fields. Thus:

$$I \propto \frac{1}{s^2}$$

where I is intensity and s the distance apart. If we define intensity as the rate of energy transfer through unit area, the diagram shows why the law holds. The area covered by the radiation increases as the square of the distance.

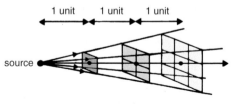

inverse square law

inverting amplifier *See* **operational amplifier.**

ion An **atom** or group of atoms that is charged because it has too many or too few **electrons** in the cloud surrounding it. The former type is a negative ion, or *anion*; the latter, which is positive, is a *cation*. **Metals** consist of ions in an electron gas; electrolytes contain anions and cations. *See also* **conduction, electrical; electrolysis.**

ionizing radiation Radiation energetic enough to ionize atoms. The radiation may be high-energy **electromagnetic radiation** (**ultraviolet, X-ray** or **gamma-ray** photons) or a beam of particles with mass (for example, alpha and beta particles from **radioactive** substances).

isobars 1. Lines on a weather map joining points of equal **air pressure**. Winds might be expected to blow across the isobars from high pressure to low. However, the rotation of the Earth makes the picture less simple, and the winds blow roughly parallel to the isobars.
 2. Nuclei of different elements with the same **nucleon number** (and therefore approximately the same mass). For example, hydrogen-3 (tritium) and helium-3 are isobars. *See also* **isotopes.**

	Protons	Neutrons	Nucleons
$^3_1$H	1	2	3
$^3_2$He	2	1	3

isotopes Species of the same chemical element, whose nuclei have the same number of **protons** but different numbers of **neutrons**. Isotopes of an element have identical chemical properties but different physical properties. A given chemical element is defined by the number of **protons** in its nuclei – this is its **proton number** Z. The proton number equals the number of electrons in the cloud of the **atom** if this is neutral. Hydrogen, the element with one proton in each of its nuclei, has three isotopes. The table shows their structure in terms of proton number Z, **neutron number** N and **nucleon number** A.

Element	Isotope	Z	N	A	Symbol	Abundance
Hydrogen	Hydrogen	1	0	1	$^1_1$H	99.98%
	Deuterium	1	1	2	$^2_1$H or $^2_1$D	0.02%
	Tritium	1	2	3	$^3_1$H or $^3_1$T	radioactive

A sample of **deuterium** has twice the mass of the same number of ^{1}H atoms, so the substance is sometimes called 'heavy hydrogen'. Tritium is three times as dense: it is radioactive, with a short **half-life**, so is not found on the Earth. *See also* **uranium.**

J

jet engine An engine in which air is drawn in, mixed with fuel and burnt, with the exhaust gases leaving at very high speed to provide propulsion. The law of conservation of **momentum** shows that a small mass of gas moving at very high speed can give a very large mass (for example, an aeroplane) a small increase in speed. Sir Frank Whittle developed the jet engine during the 1930s.

jet stream A narrow band of very fast wind that snakes around the Earth from west to east at a height of between 10 and 16 km above ground. The two main jet streams occur at about latitudes 40° to 50° north and south, above the boundaries between cold polar air and warm tropical air. The jet stream travels at over 150 km per hour. It is used to assist aircraft in crossing from west to east, and it influences the weather by controlling the warm air/cold air boundary whose mixing produces 'lows'.

joule (J) The unit of energy and work. A joule is the work done when a force of 1 newton moves an object through a distance of 1 metre.

The joule is named after James Joule (1818–89), a British physicist whose major work helped to provide a better understanding of the nature of energy.

joule meter A device used to measure the energy transfer in an electric circuit. The meters used in houses to display the electrical energy taken from the mains are of this type.

K

keeper A soft-iron or steel bar placed across the **poles** of a permanent **magnet** to maintain its strength when it is not in use. It maintains the closed chain of **domains**.

Kelvin scale The scientific (or *absolute*) scale of temperature, which begins at absolute zero (0 K). The size of the unit (the kelvin, K) is exactly the same as that of the Celsius degree (°C). The melting-point of ice is 273.15 K. Convert Celsius temperatures to Kelvin temperatures by adding 273.15:

$$0°C = 273.15 \text{ K}$$
$$100°C = 373.15 \text{ K}$$

The Kelvin scale is used in the **gas laws** formulae. It is named after William Thomson, Lord Kelvin (1824–1907), a British physicist.

Kepler's laws Three laws of planetary motion discovered by the German astronomer, Johannes Kepler (1571–1630). They are:

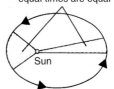

areas covered in equal times are equal

* The orbit of each planet is an ellipse with the Sun at one focus.
* The radius vector joining each planet to the Sun covers equal areas in equal times.
* The square of the orbital period ('year') of a planet is proportional to the cube of its mean distance from the Sun.

Kepler's laws
Note that the actual orbits are much less elliptical than the diagram suggests.

These laws were deduced from careful observations taken over many years. Later, Newton showed that they were a natural consequence of his laws of force and of gravity.

kilowatt-hour (kWh) The unit of energy in which electricity is expressed by suppliers. It is the energy transferred by an electric device rated at 1 kW in 1 hour:

$$1 \text{ kWh} = 1000\text{W} \times 60 \times 60 \text{ s} = 3.6 \text{ MJ}$$

kinetic energy The **energy** of motion. The kinetic energy of a moving object is the work needed to stop it. The value depends on the object's mass m and on its speed v as follows:

$$E = \frac{1}{2}mv^2$$

Its unit is the **joule** (J). Thus, the kinetic energy of a 10-tonne spacecraft in orbit at 8 km/s is

$$\tfrac{1}{2} \times 10{,}000 \times (8000)^2 \text{ J} = 320 \text{ GJ}$$

kinetic model or kinetic theory The generally accepted theory of the structure of **matter**. Four statements summarize it:
• All matter consists of particles.
• The particles are always moving.
• The particles tend to attract each other.
• Temperature relates to the mean energy of the particles.
The evidence for this model is its success in explaining how matter behaves. *See*, for instance, **adhesion, Brownian motion, conduction, convection, diffusion, evaporation, expansion, gas laws, state, state change**.

L

laser A device able to produce an intense, narrow beam of radiation. The name stands for Light Amplification by Stimulated Emission of Radiation.

Energy is absorbed by the atoms of the substance used. These release the energy in the form of photons. The light waves travel in the same direction, with crests aligned with crests and troughs with troughs (i.e. in **phase**) and all with the same frequency. As a result, the output beam is powerful, in phase and monochromatic (of one colour).

Lasers are used in **fibre optics**, CD players, **holography**, electronic engineering, science and industry. They are also useful for teaching the properties of waves.

latch *See* **bistable**.

latent heat The energy involved in **state change**. 'Latent' means 'hidden'. If energy is added at a constant rate to a pure solid sample, the temperature/time graph is as in diagram (a). A liquid losing energy at constant rate will fall in temperature as in (b).

In each case the temperature stays constant while the change of state takes place. All the same, there is still energy transfer to or from the particles. The quantity of energy transferred during state change depends on what the substance is and its state. *See also* **specific latent heat**.

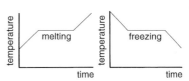

latent heat *Temperature remains constant during state changes between (a) solid to liquid; (b) liquid to solid.*

lateral inversion The apparent left-to-right reversal of an image behind a mirror. The effect follows from the laws of **reflection**.

law (in science) A statement that describes the general behaviour of a physical effect or the relationship between effects or variable quantities. Some laws are very restricted: for example, **Hooke's law**, which applies only to certain elastic materials within a limited range of applied force, and **Ohm's law**, which applies only to some conductors under certain special circumstances of use. Other laws are more fundamental, such as the **law of conservation of energy**, and **Newton's laws of motion**, and are believed to apply very widely, if not universally.

Essentially, all scientific laws are generalizations (or models) based on experimental observation, and cannot be called 'true' because they may be disproved by new experiments or observations. **Newton**'s law of gravity, for example, was accepted for 200 years, but was then seen to be of restricted application with the advent of **relativity**. While Newton's law is true on a local scale, it is really only a special case of **Einstein**'s law of gravity. That in turn may need to change in the future.

LCD *See* **liquid crystal.**

LDR *See* **light-dependent resistor.**

LED *See* **light-emitting diode.**

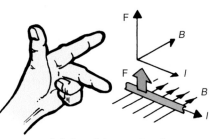

left-hand (motor) rule A rule that helps the user to recall which way the **motor effect** acts. Set the thumb and first two fingers of the left hand at right angles to each other. They then

left-hand (motor) rule

show the directions of the motion: **f**irst finger = magnetic **f**ield, *B*; se**c**ond finger = **c**urrent, *I*; thu**m**b = **m**otion (in direction of force, *F*).

lens An optical element that bends light by **refraction.** *Converging* or *positive lenses* bring rays together; *diverging* or *negative lenses* spread them out.

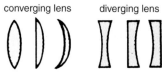

converging lens   diverging lens

lens *Common shapes of converging and diverging lenses.*

Glass converging lenses used in air are thicker in the centre than at the edges; diverging lenses are thinner. *See also* **focal distance.**

Ray diagrams are used to find where the image of an object will appear after refraction by a lens.

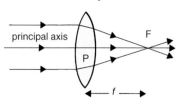 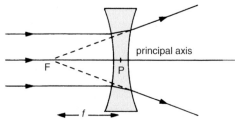

lens *The refraction of light by a converging lens.*

lens *The refraction of light by a diverging lens.*

In the ray diagrams shown,

(a) The **pole** P is the centre of the lens.

(b) The *principal axis* is the *normal* (perpendicular) through P.

(c) The *focal point* F is the point through which light that is incident (incoming) close to the axis and parallel to it passes or seems to pass after refraction. FP is the focal distance, *f*.

(d) The **object** O is the actual source of the rays concerned.

(e) The **image** I is located where the rays seem to come from.

Lenses are used in many **optical instruments**, in order, for example, to produce an image in a certain place, or to produce an image larger than the object. Contact lenses and the lenses in glasses can correct **vision defects**.

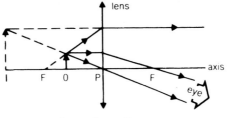

lens *A ray diagram.*

Lenz's law A law stating that the direction of the induced voltage in **electromagnetic induction** is such that it opposes the change causing it. For example, when a conductor moves in a magnetic field the voltage induced produces a current in the conductor. The **motor effect** on this current due to the magnetic field produces a force acting in the opposite direction to the motion that produced the voltage in the first place.

When an induced voltage is produced by a changing current, as in a transformer, for example, the induced voltage acts in the opposite direction to the change in current – tending to reduce the current when it is increasing, and tending to increase the current when it is decreasing.

Lenz's law is a consequence of the law that energy must be conserved, i.e. energy cannot be obtained from an induced voltage without doing some work against a force. *See also* **energy, law of conservation of**.

The law is named after Heinrich Lenz (1804–65), a German physicist who worked most of his life in Russia. He explored electrical conductivity and the heating effect of current, as well as induction.

lepton *See* **particle physics**.

lever A type of simple **machine**, in which a certain force applied at one point on a rod gives an output force elsewhere. Its action is based on the law of **moments**. It is common to group levers into three classes (*see diagram*). They differ in the positions of the input and output forces (*the effort* applied and the *load* overcome) relative to the pivot (turning point or *fulcrum*).

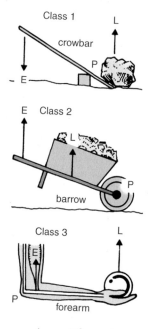

Class 1

crowbar

The pivot is between the load and the effort. Usually the effort is smaller than the load because it is further out.

Class 2

barrow

The pivot is at the end of the lever and the load is in the middle. A small effort lifts a large load.

Class 3

forearm

The pivot is again at the end but the effort is in the middle. There is a mechanical disadvantage and a large effort is needed to lift a small load.

lever *The main types of lever and how they operate.*

light Electromagnetic waves with wavelengths between 400 and 760 nm, that the normal human eye can detect. Like all waves, light can be absorbed, reflected, refracted, diffracted and show **interference** effects. Like all electromagnetic waves, it travels through empty space at 300,000 km/s and can be polarized. *See also* **diffraction, eye, lens, optical instruments, polarization, reflection, refraction, spectrum, speed of light.**

light-dependent resistor (LDR) A resistive device whose resistance is affected by light shone on it. LDRs are made from a semiconductor material (usually cadmium sulphide) in which electrons are freed to act as conductors when they gain energy from light. A typical LDR has a resistance of several million ohms (megohms) in the dark but this decreases to a few kilohms in bright light. They can be used to measure light intensity (e.g. in photometers), and are used in control circuits

Light-dependent resistor symbol

which, for example, switch on an outside light when it gets dark, or adjust the aperture in a camera for different light levels.

light-emitting diode (LED) A p–n junction **diode** usually made from gallium arsenide phosphide (*see* **semiconductor**). The recombination of **holes** and **electrons** releases energy which appears as light. The junction is made near the surface so that the emitted light can be seen. LEDs are used in electronic displays.

light-emitting diode symbol

lightning The large spark that jumps between ground and cloud (and vice versa) in a thunderstorm. A very high voltage can appear between the top of a thundercloud and its base, and between the base and the ground. If the voltage is high enough, perhaps 100 MV, a lightning **discharge** may occur. A current of 100 A may exist for 0.1 s, with an energy output in the form of light and heat of 10^9 J. The sudden expansion of the air causes the crack of thunder.

No *lightning* rod could carry the huge energy of this discharge. In fact lightning rods discharge electricity from the ground and so *prevent* lightning. *See also* **charge distribution**.

light-year A unit of distance used in popular astronomy. It is the distance travelled by light in one year (about 10^{16} m).

line of force An imaginary line that shows the *direction* of a force field. In magnetism the direction is that of the force on a north-seeking pole; in electricity it is the direction of the force on a positive charge. *See* **electric field**, **magnetic field**.

liquid crystal A substance that flows like a **liquid** but in which groups of **molecules** can be aligned as in a **crystal**. In an ordinary liquid the particles move around at random: there is no structure. On the other hand, in a solid crystal, the particles are fixed in a very regular pattern.

material in solid state

liquid crystal

material in true liquid state

liquid crystal *Molecular structure of a substance in solid, liquid crystal and liquid states, showing an increasing degree of disorder.*

The particles of a 'liquid crystal', which is in fact a liquid, are long and thin and tend to line up one way. An outside electric field affects the way they line up and reflect light; this feature is the basis of the *liquid crystal display (lcd)*. The substance is contained between two conducting glass plates; if there is a field between the plates, the particles line up one way – if not, they line up a different way. In one state the liquid blocks light, in the other it allows the light to pass.

lithosphere The outer part of the Earth, comprising the crust and the upper part of the mantle. It consists of large plates in motion relative to each other. *See* **plate tectonics**.

litre   A unit of volume, symbol l. It is widely used but does not belong to the SI system of units (*see Appendix A*). The name comes from the Greek for 'pound'. The litre is the volume of a kilogram of pure water at 4°C. The SI unit of volume, the cubic metre (m³), is equal to 1000 l.

live wire The wire in domestic circuits that maintains a high **potential difference** compared with the neutral or zero voltage. In Europe its **insulation** is coloured brown and this is the wire that should be fused. *See also* **mains plug**.

logic gate *See* **gate**.

longitudinal wave *See* **wave**.

long sight *See* **vision defects**.

loudness A measure of the perceived intensity of **sound** waves. Perceived loudness is different from sound intensity because the ear is more sensitive to some frequencies than to others.
  Loudness is measured in dB(A), intensity in dB (*see* **decibel**, **intensity**). A loudness of more than 100 dB(A) is very loud; loudness of more than 120 dB(A) (as in some discos and factories, or near a road drill) can damage hearing.

loudspeaker A device that uses a varying electric current to produce a sound; i.e. a type of **transducer**. In one design, a varying current in a coil of wire produces a varying magnetic field. This causes

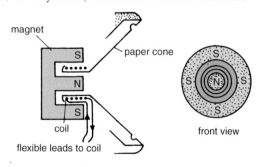

loudspeaker *A moving-coil loudspeaker.*

the coil to move relative to a fixed circular permanent magnet. The moving coil in turn moves a paper cone and produces sound.

lubrication The use of a substance to reduce **friction** between surfaces. Oil and graphite dust are common lubricants. They hold the surfaces apart and thus reduce **adhesion**.

M

machine A device in which a force applied at one point gives an output force elsewhere. Simple examples are **levers**, **pulleys** and **hydraulic machines**.

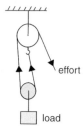

In most cases the output force (the *load* overcome) is greater than the input force (the *effort* applied): the **force ratio** (or *mechanical advantage*) is greater than 1. This is balanced by the need for the input force to move further than the output: the **distance ratio** (or *velocity ratio*) is more than 1. The efficiency of a machine is the fraction below expressed as a percentage.

machine *A simple two-pulley system with a distance ratio (velocity ratio) of 2.*

$$\frac{\text{output work}}{\text{input work}} \quad \text{or} \quad \frac{\text{work done on load}}{\text{work done by effort}}$$

Work has to be done against friction and in lifting 'non-load' weights such as ropes and pulleys, so that the efficiency is usually much less than 100%.

magnet An object made of a ferromagnetic material such as iron, with small regions called **domains** with individual magnetic fields that are aligned in the same direction. Thus the magnet has **poles** and a **magnetic field** around it. Magnets or **electromagnets** (whose effect can be switched on and off) are used in many electric machines.

The *law of magnetic force* describes how poles affect each other: like poles repel each other; unlike poles attract each other.

magnetic field A region of space in which there is magnetic force. Such **fields** are three-dimensional; in most diagrams, sections are shown. The diagrams (*shown opposite*) use *field lines* or **lines of force** to show the field direction and strength (represented by the closeness of lines) at each point. A field line is defined as the path a free N-pole (if there were such a thing) would follow. *See also* **Earth's magnetism**, **magnetic material**.

magnetic material A substance that experiences a force when in a **magnetic field**. Iron, cobalt, nickel can be strongly magnetized; they are *ferromagnetic*. Even stronger magnets can be made from oxides of iron called *ferrites*. Magnets made from materials that are difficult to magnetize

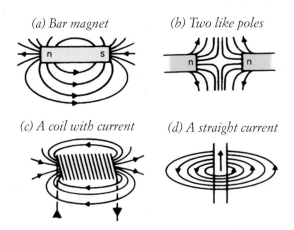

(a) Bar magnet *(b) Two like poles*

(c) A coil with current *(d) A straight current*

magnetic field *Sections of some simple magnetic fields.*

and demagnetize are called *permanent magnets* (sometimes said to be made from 'hard' magnetic materials). Magnets which are easily formed and easily demagnetized are *temporary magnets* (made from 'soft' materials, e.g. 'soft' iron). Permanent magnets are used for loudspeakers, compasses etc. and temporary magnets are used for transformers, electromagnets, etc.

magnification The apparent increase in size of an image compared with the **object**. It is often defined as the image height (distance from the axis) divided by the object height. Where the optical system is a single **lens** or mirror then

$$\text{magnification} = \frac{\text{image height}}{\text{object height}} = \frac{\text{image distance}}{\text{object distance}}$$

where the image and object distances are measured from the **pole**. *See* **ray diagram**.

magnifying lens or magnifying glass A single converging **lens** used to produce a large, upright, virtual **image** of a small object. The object must be between the **pole** and focal point of the lens. This is also called a simple **microscope**.

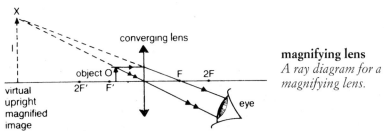

magnifying lens
A ray diagram for a magnifying lens.

main sequence *See* **Herzsprung–Russell diagram**.

mains plug A device used to join the cable of an electrical appliance to a mains supply socket. In the UK, mains cable has three wires: *live* (brown coat), which carries the **current**; *neutral* (blue), for its return; and *earth* (green/yellow) for safety. Each wire goes to a pin in the plug as shown in the diagram.

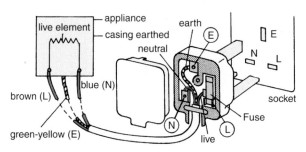

mains plug

manometer A device used to measure **pressure**. Examples are shown in the diagram (*see opposite*). *See also* **barometer, Bourdon gauge**.

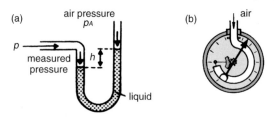

manometer *Common manometers:*
(a) liquid manometer; (b) a Bourdon gauge.

mass A measure of the quantity of matter in an object; the unit is the kilogram (kg). It depends on the number of **molecules** and their masses and does not depend on gravity. For this reason, an object would have the same mass on the Earth and on the Moon but a different weight because of gravity change. A **balance** that is not affected by variations in gravity is used to measure mass.

mass-energy The relationship between an object's mass and energy, according to Einstein's theory of **relativity**; if the object gains energy, its mass increases, and vice-versa.

The equation that relates the mass change Δm to the energy change ΔW is $\Delta W = \Delta mc^2$. Here c is the speed of light in empty space (3×10^8 m/s).

Thus, if a reaction converts 1 mg of mass to energy, the energy transfer is 9×10^{10} J.

Nuclear power depends on the energy released when atomic nuclei are altered into lower-mass forms.

The *law of constant mass-energy* is that the sum of an object's mass and energy is constant unless energy or mass is transferred to or from it.

mass number *See* **nucleon number**.

matter Anything that takes up space, especially a physical material: anything not a **vacuum**. Normal matter can exist in any of three main **states**: gas, liquid, or solid. Matter consists of very small particles. *See also* **atom, kinetic model, molecule**.

mean One type of **average**. To find the mean of a set of values, add them and divide by the number of values in the set. *See* **median, mode**.

mechanical advantage An old name for **force ratio**:

$$\text{Force ratio} = \frac{\text{output force}}{\text{input force}} = \frac{\text{load}}{\text{effort}}$$

mechanics The branch of physics concerned with **forces** and their effects on whole objects. **Statics** deals with balanced forces; **dynamics** relates forces to the change of motion produced; *kinematics* describes the motion of objects (*see* **equations of motion**).

median One type of **average**. The median of a set of values is the one in the middle, when the values are in order. **See also mean, mode**.

medium Any substance through which a form of **radiation** travels. In this context, a medium's main property is the speed of the radiation through it. *See also* **speed of light, speed of sound**.

melting temperature or melting point The temperature at which a pure solid turns to a liquid or the liquid turns back to the solid, i.e. the temperature at which the **state** changes. *See* **latent heat, state change**.

meniscus The curved surface of a liquid in a tube. It is an effect of **surface tension**. *See also* **capillarity**.

meson *See* **particle physics**.

metal Any of a class of chemical elements with the following properties:

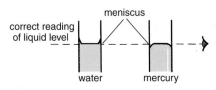

meniscus *The menisci of water and mercury.*

physical: Metals are shiny and good thermal and electrical conductor; when solid, they are malleable (can be hammered into thin sheets) and ductile (can be drawn out into thin threads);

chemical: Metals make positive ions in solution or in reactions; they replace hydrogen in acids to form salts, and they form alloys.

Metals are held together by the *metallic bond* which is due to metal atoms losing electrons surrounding the ions; these electrons move around at random as an 'electron gas'. It is the free electrons that make metals good thermal and electrical conductors. The atoms are arranged in regular lattices forming crystals, but a typical metal object is made of a large number of small crystals (i.e. it is *polycrystalline*). Metals tend to be easy to liquefy (mercury is a liquid at room temperature). Most metals occur naturally in the Earth's crust in compounds called *ores*, from which they have to be extracted and purified.

⊕ positive ion

➢ free electron

metal *A 'gas' of electrons surrounds the positively charged ions.*

meteorology The study of climate and weather patterns, with the aim of trying to explain, predict and change them.

meter A measuring device. Thus a force meter measures **force**, a thermometer measures **temperature**, and an ammeter measures **current**. *See also* **voltmeter**.

method of mixtures A technique that allows one to measure such properties of matter as **specific thermal capacity**. It involves mixing samples of known masses that differ in temperature, and finding the equilibrium temperature. There should be no energy transfer in or out of the mixture.

microelectronics The use of **integrated circuits** (microchips) for tasks in **information technology** and **control**. Microchips are cheap to build and run, robust, and very small. They are, as a result, widely used in modern audio, video and computing equipment.

microphone A device that converts **sound** waves to a changing electric **current**. The current may be amplified and passed to an oscilloscope, recorder or speaker. A moving-coil microphone is similar to a **loudspeaker** but works in reverse.

For a diagram of the simple carbon microphone, *see* **telephone**.

microscope An **optical instrument** used to give a large image of a small object. The *simple microscope* is a single converging **lens** (*see* **magnifying lens**). *Compound microscopes* have two or more lenses; that nearest the object is the *object lens* (objective), while that nearest the eye is the *eye lens* (eyepiece).

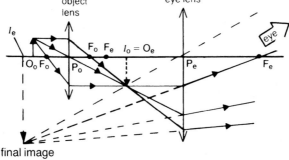

microscope *A ray diagram of a compound microscope.*

microwaves A region of the **spectrum** of **electromagnetic waves**, in the wavelength range from about 1 mm to 300 mm, with a frequency range of 1000 MHz to 300,000 MHz. (Definitions of the frequency range vary.) Microwave sources include tubes called klystrons. The uses of microwaves include **radar**, microwave ovens (the wavelengths used being strongly absorbed by the hydrogen atoms in the water contained in foods), and telephone communications. Microwave sources and detectors are useful in studying the behaviour of waves.

mirage An optical illusion caused by **refraction** in air of non-uniform temperature. It is seen when the air close to the ground is warmer than that higher up. As light nears the ground, it bends away, and an image appears on the other side of the surface (*see diagram*). An image of the sky gives the effect of a pool of water. *See also* **total internal reflection**.

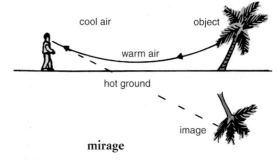

mirage

mirror An optical device that alters the path of light by **reflection**. When parallel beams of light strike a flat (plane) mirror at any angle, the reflected beams are parallel. The image appears to be the same distance behind a flat mirror as the object is in front, and is the same size and the same way up. *See diagram 1.*

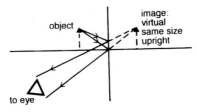

mirror *(1) How a flat (plane) mirror forms an image.*

A *diverging mirror*, such as a driving mirror, has a convex surface (like the outside of a bowl). Rays that are parallel when they strike it spread out after being reflected. A *converging mirror* has a concave surface (like the inside of a bowl). When parallel rays fall on it, the reflected rays come together. *See diagram 2.*

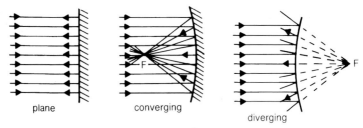

mirror *(2) Types of mirror.*

A **ray diagram** can be used to find the image of an object formed by a mirror– *see diagram 3 overleaf,* where *F* is the focal point of the mirror (*see* **focal distance**). The object must be closer than *F* to the mirror to form a *real* image (*see* **image**).

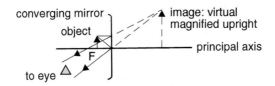

mirror *(3) Ray diagram of a converging mirror (make-up or shaving mirror).*

mode One type of **average**, the others being **median** and **mean**. The mode of a set of values is the most common value.

model A simplified representation or description of a complex system, such as the universe, designed to help understanding of it. In science, a model may not be a visual picture at all: many models are purely mathematical patterns. On the other hand, sometimes a model is just another name for a **theory**. Thus we can speak of the atomic theory or the atomic model of matter.

Mohs' scale *See* **hardness.**

molecule The smallest quantity of a chemical substance that can exist and take part in a chemical reaction. It may consist of a single **atom** or a group of atoms bonded together. Molecules of many gases consist of two identical atoms bound together, e.g. oxygen O_2, hydrogen H_2; other simple molecules are ammonia NH_3, and water H_2O. Organic substances and plastics are usually large molecules with large numbers of atoms – glucose has 24 atoms, haemoglobin has many thousands. Plastics are long chain molecules with an arbitrary number of atoms in a repeating pattern, as is the DNA molecule.

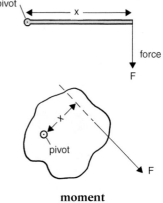

Note that some everyday substances do not exist as molecules, e.g. common salt (NaCl), which in the solid state consists of a regular array of ions and in solution or as a liquid consists of separate ions.

moment or torque The turning effect of a force or **couple** around some point. The moment of a force is the product of the force F and its perpendicular distance x from the turning point: $T = Fx$ The unit is the newton-metre, Nm.

moment

momentum The product of an object's **mass** and **velocity**. The unit of this **vector** is the kilogram metre per second, kgm/s (or newton-second, Ns).

The *law of conservation of momentum* follows from **Newton's laws of motion**. It states that the total momentum of a system of objects is constant unless a net outside force acts. The law applies to cases of impact, as in diagram (a) overleaf, and of disintegration, (b). All directions must be signed, e.g. to the right as positive, to the left as negative, so u_2 is negative:

$$m_1u_1 + m_2u_2 = m_1v_1 + m_2v_2$$

In each case the total momentum of the system is the same after as before the event. (*See diagram overleaf.*)

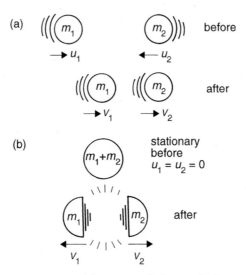

momentum *Momentum before and after*
(a) impact and (b) explosion.

monitor 1. To keep a close watch on something, or collect data continuously or at regular intervals.

2. The display device of a computer or TV system.

Moon The Earth's natural **satellite**; its **gravity** is about a sixth of that of Earth. That is not enough for the Moon to have any atmosphere, so there is no known life and no water. The Moon travels round the Earth in about 28 days (a month); it turns on its axis once in the same time which means that only one side can be seen from Earth. During the month, a cycle of *Moon phases* is seen, from full Moon (a full bright disc) when the Moon is the other side of us from the Sun, to new Moon (nothing at all) when the Moon is between us and the Sun, and back again. *See also* **eclipse**.

The Moon has the following statistics: mass 7.35×10^{22} kg (0.012 Earth mass); radius 1738 km (0.27 Earth radius); density 3340 kg/m^3; surface gravity 1.6 N/kg; mean distance from Earth 384,400 km.

motor effect The **force** on a current in a magnetic field. Magnetic fields are used in some types of **cathode-ray tube** to deflect the current of electrons and move the spot on the screen. If the current

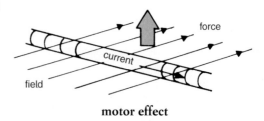

motor effect

is in a wire, the wire itself may move. Use the **left-hand (motor) rule** for the direction of motion.

The effect is the basis of the action of the moving-coil **loudspeaker** and **ammeter** and the electric **motor**. In each case the force causes a coil to move in a field.

motor, electric A device which uses the interaction between an electric current and a magnetic field to produce motion. In most cases a central coil rotates. The structure of a simple motor is just like that of a **generator** (*see diagram at that entry*). Direct current enters the coil through carbon brushes pressing on a split ring; in the simple ac motor, the carbon brushes press on two slip rings. As the motor is just like a generator, it in fact acts as a generator. A back-emf appears in the turning coil; this tends to oppose the applied voltage. *See also* **electromagnetic induction, motor effect**.

music Sounds arranged in such ways that they can bring pleasure. Each sound (*note*) can be described in terms of the following:
(a) pitch: this relates to **frequency**;
(b) loudness: this relates to **amplitude**, energy content;
(c) timbre, or quality of sound: this relates to the mixture of different
    frequencies appearing in the sound.

During the course of history, a number of types of music have appeared. These depend on different scales and on different views of which notes sound pleasant together. *Compare* **noise**.

N

NAND gate *See* **gate**.

natural frequency The frequency at which any system that can show **vibration** will do so most freely. The value of this depends on the physical nature of the system.

Thus, if pushed, a simple **pendulum** will swing at a frequency that depends on its length and on the gravitational field strength at that place. The frequency of the note produced by blowing over the top of a bottle depends on the size and shape of the space inside. **Resonance** occurs when an object capable of vibration is forced or pushed regularly at the same frequency as its natural one.

neutrino A **fundamental particle** of matter, carrying no charge and with a very small (but probably non-zero) mass. The existence of the neutrino was predicted in 1930 to explain why **beta particles** were emitted in radioactive decay with a range of energies: the neutrino carries away some of the energy released in the decay. The neutrino was first observed experimentally in 1955.

neutron One of the three main atomic **particles** found in the atomic nucleus. The neutron has no charge (hence its name); on the atomic scale its mass is approximately 1 **atomic mass unit**. *See also* **nucleon number**.

A nucleus of every atom except that of common hydrogen, contains one or more neutrons. Light nuclei have roughly as many neutrons as **protons**; heavy ones may have about 1.5 times as many. These figures are only approximate – the value of **neutron number** N can vary quite widely for a given **proton number** Z. The result is the set of **isotopes** of each element.

Outside the nucleus the neutron is not stable; it decays into a proton and an electron, or **beta particle**. The **half-life** of this decay is about 1000 s.

neutron number (N) The number of neutrons in the **nucleus** of an atom, i.e. the **nucleon number** minus the **proton number**.

newton (N) The unit of **force**. One newton is the force needed to accelerate one kilogram by one metre per second per second.

Newton, Isaac (1642–1727) An English scientist of outstanding importance. His main work was in mathematics: it was the mathematics of the motion of planets and moons that led him to his theory of **gravity**. The theory was important in linking the motion of objects on Earth and that of objects in the sky. His work in physics included the study of light, gravity, statics and dynamics.

Newton's laws of motion or laws of force Three statements which are the basis of Newtonian (everyday) mechanics. They define a force in terms of its effects on objects, and relate forces to each other. The laws state that:

(1) an object will remain at rest (still), or if moving will continue to move at a constant speed in the same direction, unless acted upon by an external force. (The force may be a single force or the resultant of several unbalanced forces.)

(2) An external resultant force alters the motion of an object in such a way that the rate of change of its **momentum** is proportional to the force and in the same direction.

This law may be expressed mathematically as:

$$\text{impulse} = \text{change of momentum } (\Delta P)$$

$$Ft = \Delta P$$

and:

$$\text{force} = \text{mass} \times \text{acceleration}$$

$$F = ma$$

(3) When two objects (A and B) interact, the force that A exerts on B is equal in strength to the force exerted by B on A, but the two forces act in opposite directions. (This means that forces always occur in equal and opposite pairs).

Law 2 is used to define the unit of force, the **newton**.

See also **centripetal force, equilibrium, inertia.**

newton meter *See* **force meter.**

nodes Places in **standing waves** at which there is zero displacement. They occur because of interference between waves travelling in opposite directions. At the nodes the displacement from one wave is always exactly equal in size but opposite in direction to the displacement due to the other wave. At points halfway between nodes there is maximum vibration as the combined displacement due to the two waves swings from a maximum negative value to a maximum positive value. These points are called *antinodes*.

In the diagram, a stretched string vibrating with three cycles of standing waves has nodes at the points N, and antinodes at the points A.

See also **sound sources.**

noise Sound that is unpleasant to hear, in most cases because it is a mixture of **frequencies** that do not relate closely to each other. (Frequencies *harmonize* if they bear simple ratios to each other.) Noise is therefore a mix of random frequencies; the term is used as such in electronics and other fields as well as in sound.

White noise contains a mix of all frequencies in a range. It is, for instance, the background hiss produced by radio and TV sets.

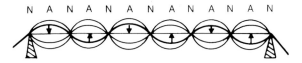

nodes *Nodes and antinodes on a vibrating stretched string.*

noise control The reduction of unwanted sound. **Noise** can be reduced by better design of the source (e.g. making quieter motors) and by **insulation**. The latter involves using something that absorbs sound waves efficiently, like foams and rubber.

NOR gate *See* **gate**.

nuclear fission *See* **fission, nuclear power**.

nuclear fusion *See* **fusion, nuclear power**.

nuclear power The supply of useful energy from nuclear reactions. A nuclear power station is built round a nuclear reactor. The only working types so far involve **fission**. **Fusion** is in theory more effective, safer and less costly, but in spite of many years of research it is not yet possible to control the process to produce a continuous supply of power.

In a fission reactor or *pile*, a controlled **chain reaction** takes place at a steady rate. The kinetic energy of the particles produced in each fission is converted into heat; a *coolant* removes this thermal energy to make steam. The steam drives turbines linked to generators.

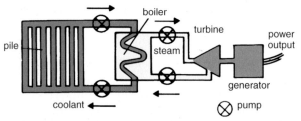

nuclear power *The operation of a nuclear power station.*

The central pile of the *thermal reactor* has two components apart from the fuel. A moderator must slow the **neutrons** made in each fission to a speed low enough to cause more fissions. Graphite is commonly used for this purpose. *Control rods*, often made of boron, absorb neutrons, limiting the rate at which fissions occur. They are raised and lowered in the core to control the reaction.

The pile is surrounded by thick shielding of steel and concrete. This prevents undue radiation from the core putting people outside in danger.

The diagram shows a standard type of fission reactor in simplified form. The fuel in this thermal reactor is enriched **uranium** – it has a high content of the active isotope ^{235}U.

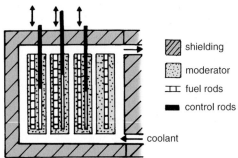

shielding

moderator

fuel rods

control rods

coolant

nuclear power *Cross-section of a typical fission reactor.*

Many people are concerned about the safety of nuclear reactors. There is, as yet, no completely safe way of disposing of radioactive waste (including the used uranium fuel rods). There have also been a number of accidents at nuclear plants that have released radioactivity into the atmosphere (the most serious of which was at Chernobyl in the Ukraine, which killed many people and caused major environmental damage).

nucleon A massive **particle** found in the nuclei of atoms, either a **proton** or a **neutron**.

nucleon number (A), atomic mass or mass number The number of **nucleons** – protons and neutrons – in the nucleus of an **atom**. As the masses of these particles are each very nearly 1 on the atomic scale, the nucleon number is also called the *atomic mass*, or *mass number*. The nucleon number A is the sum of the **proton number** Z and the **neutron number** N, as in the tables. The number of electrons in a neutral atom equals the proton number.

Particle	Z	N	A
electron	–	–	0
neutron	0	1	1
proton	1	0	1
heliumatom	2	2	4

The **isotopes** of an element have the same proton number but differ in nucleon number. This is because their nuclei do not have the same numbers of neutrons. On the other hand **isobars** have the same value of A, but differ in N and Z.

Element	Species	Z	N	A	
hydrogen	hydrogen	1	0	1	
	deuterium	1	1	2	} isotopes
	tritium	1	2	3	} isobars
helium	helium-3	2	1	3	
	helium-4	2	2	4	} isotopes
	helium-6	2	4	6	

All these species with particular proton/neutron combinations are called *nuclides*.

In writing about atoms, nuclei, and nuclear reactions between them, three-part symbols are used. The name of the element is given by the standard chemical symbol. Before it, above the line, is the value of A; before it, below the line, is the value of Z. Thus the six species in the accompanying table are written as follows:

$$^1_1H \quad ^2_1H \quad ^3_1H \quad ^3_2He \quad ^4_2He \quad ^6_2He$$

$^{235}_{92}U$ is an isotope of the metal **uranium**, in fact the one whose nucleus has 92 protons (as all nuclei of uranium have) and 143 (= 235 – 92) neutrons. The neutral atom of uranium-235 has 92 electrons in the cloud round the nucleus.

nucleus The compact, relatively massive core of an **atom**. It carries a positive charge because it contains positive **protons**. The mass depends on the number of **neutrons** and protons (*see* **nucleon number**).

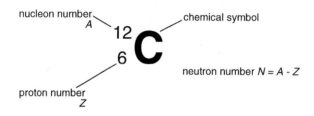

nucleus

The nuclei of atoms and most **ions** have clouds of **electrons** around them. Electrons carry negative charge; in the neutral atom there are as many electrons in the cloud as protons in the nucleus.

A nucleus is about 10^{-14} m across. This makes it far smaller than an atom – the radius of electron clouds is around a thousand times bigger. The electron cloud is involved in chemical reactions; the nucleus is crucial to the physical properties of a substance. The nucleus is also the site of nuclear reactions. These can release very large amounts of energy; *see* **atomic bomb**, **nuclear power**, **radioactivity**. An atomic nucleus is usually described by a symbol like the one shown for the most common isotope of carbon (carbon-12) in the diagram.

O

object 1. A body or entity acted upon in a physical process.
2. (in optics) The body of which an optical system produces an image.
The object must be either a source of radiation or capable of reflecting it.
In complicated systems the object of one part may in fact be an image
formed by another part. *See also* **lens, mirror, telescope, microscope**.

ohm (Ω) The unit of electric **resistance**. It is the resistance between two
points on a conductor when a constant **potential difference** of 1 volt
between them produces a **current** of 1 ampere.

Ohm's law The law that states that the **current** I in a conductor at
constant temperature is proportional to the **voltage** V between its ends:
$I \propto V$. A sample's **resistance** R is defined from this: $R = V/I$. Although
Ohm's law applies mainly to metals (*ohmic conductors*) the concept of
resistance is used much more widely. *See also* **electromotive force**.
    The law is named after Georg Ohm (1787–1854), a German physicist
who first published the concept in 1827. He also explored sound waves
and the interference of light.

opaque *See* **transparent**.

open cluster A loose grouping of about 100 to 1000 stars in the same
region of space. Stars in open clusters tend to be young, and are moving
apart. The *Pleiades* is an open cluster, consisting of about 100 stars in a
space 10 light-years across. The six brightest stars are visible to the
unaided eye. *Compare* **globular cluster**.

operational amplifier or op-amp An **integrated circuit** used to
increase the amplitude of a small alternating voltage without distortion.
The name comes from the fact that the circuits were originally designed to
carry out mathematical operations and hence solve equations.
    The op-amp has two input terminals, and the amplification is of the
voltage difference between the two inputs; it is a differential amplifier.

$$\text{amplification} = \frac{\text{output voltage}}{\text{input voltage difference}}$$

The inverting input gives an amplified output but with a phase change,
i.e. a positive input gives an amplified negative output. When part of this
input goes back to the input as *negative feedback*, the amplification (or
gain) is reduced; however, it has the advantage of being more accurately

predictable and more constant for a wide frequency band. This is the use as an *inverting amplifier*.

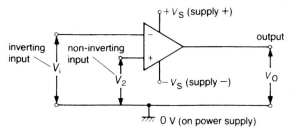

+V_S (supply +)

inverting input
V_i

non-inverting input
V_2

− V_S (supply −)

output

V_O

0 V (on power supply)

operational amplifier

optical density   The ability of a transparent medium to refract light, which is related to the speed of light in the medium. The greater the optical density of a medium the more slowly light travels through it, and the greater the **refraction** that is produced.

optical instruments   Lenses or **mirrors**, or combinations of them, used to produce **images** for some purpose. A projector, for example, focuses on a screen a real image of a small object, such as a slide or a microbe. *See also* **camera**, **eye**, **microscope**, **periscope**, **slide projector**, **telescope**.

orbit   The path, usually closed, of an object moving in a central **force** field, such as the Earth's gravitational field or the electric field round a **charge**. The **centripetal force** prevents the object from moving in a straight line. The object has **potential energy** and **kinetic energy**.
   For a spacecraft to be put into Earth orbit, it must reach a speed of about 8 km/s. It can then move in a circular orbit. A higher speed will produce an elliptical orbit. A speed of 11 km/s (the **escape speed**) will put the craft in a parabolic path: it will never return.

ore   Rock from which it is worth the effort to extract a mineral (such as oil or a metal). Extraction often involves large amounts of effort and expense and can cause **pollution**, so is not always worth doing (even if the price of the product is high).

OR gate   *See* **gate**.

oscillator   1. An object that vibrates or produces vibrations in other objects.
   2. A circuit whose output is an **alternating current**. It is normal to be able to tune the frequency of the output by changing the value of a **capacitor** in the circuit.

The active element in most oscillators is a **transistor**. This acts as an **amplifier**; however, part of the output energy returns to the input to guide the action. This is called *feedback*.

overtone A note from a **sound source** of higher frequency than the *fundamental*, the system's lowest **natural frequency**. The frequencies of fundamental and overtones are simply related to each other.

Each sound source produces overtones of different frequency and amplitude. The resultant wave (*see* **interference**) is unique to the source, and differs from that of other sources in quality or timbre.

ozone layer A region of the Earth's atmosphere with a high fraction of ozone gas (O_3, a rare form of oxygen) about 35 km above the Earth's surface. Ozone is poisonous; however, high in the atmosphere it is beneficial to life on Earth as it absorbs dangerous **ultraviolet** rays from the Sun: these cause skin cancer and other problems. If the concentration of ozone decreases, the ozone layer is described as having a 'hole', which can let those rays through. At the moment, two holes form in the ozone layer at certain times of the year as a result of **pollution**; they are above the Arctic and Antarctic only, but threaten the health of people in Australia, South America and the far north of Europe and Alaska.

P

parallax The apparent change in position of an object when viewed from different places. The most common everyday example is when trees viewed from a moving vehicle appear to shift relative to background objects such as distant hills. The difference in the angle of view to an object from two points a known distance apart (the *baseline*) can be used to measure the distance of the object. This method is used in surveying and in measuring the distance of nearer stars using the diameter of the Earth's orbit as a baseline.

parallel In electrical technology, describing components joined in a circuit so that the current splits between them. The **voltages** between the ends of the components are the same. In diagram (a) three resistors are connected in parallel. The **resistance** R of the group is given by

$$1/R = 1/R_1 + 1/R_2 + 1/R_3$$

Diagram (b) shows three parallel **capacitors**. The total capacitance C is $C_1 + C_2 + C_3$.
Compare **series**.

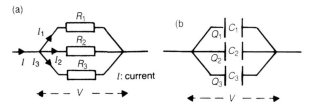

parallel (*a*) *Parallel resistors;* (*b*) *parallel capacitors.*

parallelogram rule The rule specifying the single **vector** (called the **resultant**) that has the same effect as two vectors (called **components**). In diagram (a) two tugboats are pulling a liner. The two forces on the ship are the tensions in the two cables. The ship can move only one way, not two; in effect it is acted on by a single force, the resultant of the tensions, as shown in (b).
    The steps to follow when using this rule are:
(a) Choose a scale. In the diagram it is 1 mm : 2 kN.
(b) Draw the two forces to scale, making the correct angle at the common point.
(c) Complete the parallelogram (dashed lines).

(d) Draw the diagonal through the starting point.
(e) Use the scale to find the size of the resultant. Note its direction as well.

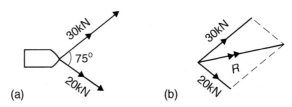

(a)　　　　　　　　　　(b)

parallelogram rule of vectors　*The resultant force is R, at 40 kN in the direction of the diagonal of the parallelogram. The liner will move in this direction.*

parsec　A unit of distance used in astronomy. It is the distance of a star whose apparent position in the sky changes by one second of arc when the Earth moves a distance of 1 AU. It is equal to 3.262 **light-years**.

particle　(in physics) An object whose volume is very small. The structure of **matter** is described in terms of particles; *see* **kinetic model**. Such particles are of various types: **ions**, **atoms** and **molecules**, these being made of smaller components (**fundamental particles**).

　In the normal atom there are three types of particle; these differ in mass and charge. In the table showing their values, atomic scales are used. The *atomic mass unit* is around 1.7×10^{-27} kg; the *electron charge* is about 1.6×10^{-19} C.

Particlce	Mass	Charge	Position in atom
electron	0	−1	Cloud
proton	1	+1	Nucleus
neutron	1	0	Nucleus

　There are hundreds of other elementary or **fundamental particles**. All particles can appear as **waves** in certain circumstances. In other words they can show wave properties like **diffraction** and **interference**. The electron microscope can be understood only by thinking of the electron beam as wave radiation; the study of this type of radiation is known as *wave mechanics*.

　On the other hand, the quantum theory states that in some cases waves must be thought of as particles. **Photoelectric effects** cannot be fully explained otherwise. In the case of electromagnetic waves they are called *photons*. In general, discrete quantities of energy associated with waves are called *quanta* (the plural of quantum). *See* **quantum physics**.

particle physics The study of the **fundamental particles** of matter. This branch of physics began with experiments on the particles emitted naturally in radioactive decay. Such experiments led to the discovery of the atomic **nucleus** (1909–11) and the **neutron** (1932). Later machines were built to accelerate charged particles and make them collide with other particles, usually the nuclei of atoms in thin metal sheets. The *linear accelerator* uses voltages (*see* **alternating current**) to give repeated 'kicks' to charged particles, accelerating them in a straight line towards a target.

The *cyclotron* uses a combination of electric and magnetic fields to accelerate charged particles by making them move in a spiral path, gaining speed and energy at each turn of the spiral.

Many new particles were discovered in *cosmic ray* studies. High-energy particles reach the Earth as *cosmic rays* (mostly high-speed **protons**), and produce other particles when they collide with nuclei in the atmosphere.

Accelerators have been developed and enlarged to accelerate protons and electrons to very high energies, equivalent to the energy gained in an accelerating voltage of a thousand billion volts. The major European research centre is in Geneva at CERN. This has the world's largest particle accelerator, a more advanced instrument than the cyclotron, called a *synchrotron*. In all modern accelerators particles are accumulated in synchrotrons called storage rings, being stored for hours before finally being accelerated and made to collide head-on.

The research has discovered a very large number of particles. Some of them are *force-carriers* (*see below*). The others are 'matter' particles, and fall into two groups, leptons and hadrons.

Leptons comprise **electrons**, muons, and tau particles, and three corresponding kinds of neutrino. These particles do not interact through the strong nuclear **force**. Only the tau has a large mass (nearly twice that of the proton or neutron).

Hadrons interact through the strong nuclear force, as well as the other fundamental forces. They are made of quarks. Mesons are medium-mass particles containing two quarks; they are unstable and so exist for only a short time after being produced in a high-energy collision. Baryons, such as the **proton** and the **neutron**, are more massive particles containing three quarks. Only the proton is stable when outside a nucleus.

Every particle has a partner of the same mass and general properties but with an opposite charge, or other basic property, called its *antiparticle*. The antiproton, for example, carries a negative charge but is otherwise the same as a proton. Antimatter like this is very rare in the universe, as far as we can tell.

Quarks are the basic particles making up the nuclei of atoms. They cannot exist on their own but combine to form nucleons, being bound together by the strong nuclear force. Leptons do not contain quarks and so do not 'feel' this force; they do not exist permanently in the nucleus, though they can be created in nuclear processes such as **beta decay**.

Particle physicists also consider forces to be carried by particles, such as the *photon, gluon, boson* and even the *graviton*.

pascal (Pa) The unit of **pressure**; the pressure of 1 newton on an area of 1 square metre, named after Blaise Pascal (1623–1662), a French scientist and philosopher who was a mathematical genius even when a boy. A major experiment involved carrying a barometer up to the top of a high mountain; Pascal showed by this that air pressure falls with height.

pendulum A regularly swinging object in which there is a regular interchange of **kinetic energy** and **potential energy**. The **period** T depends only on the length l and the acceleration of free fall g.

$$T = 2\pi(l/g)^{-1}$$

A simple pendulum has, in effect, all its mass in a small bob.

period (*T*) The time for one full cycle of a repeated (cyclic, periodic or harmonic) motion, such as the Earth's orbit round the Sun, the swing of a **pendulum**, the motion of a piston in a motor, the vibration of a guitar string. The unit is the second, s. The period of a motion is the reciprocal of its **frequency**.

When motion is shown as a graph of **displacement** against time, T is as shown in the diagram. *See also* **wave**.

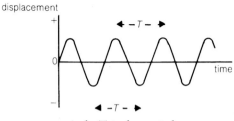

period　T *is the period.*

periscope An **optical instrument** used to give an image of an object that is hidden from view by an obstacle. A simple periscope (*see opposite*), (a), has two mirrors set in a tube. Prisms, (b), give a brighter image, but are more costly. Lenses may be added to give magnification. *See also* **total internal reflection**.

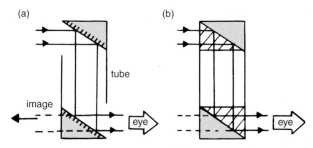

Periscopes *using (a) mirrors and (b) prisms to deflect light.*

phase The stage reached by a **wave, vibration** or other cyclic process at a given moment. As all waves and vibrations can be shown by sine curves, the phase is the point on that curve at a particular moment.

The idea is often met in the context of a *phase difference*, the difference of phase between two otherwise identical waves at a point. The waves in diagram (a) have a quarter-wave phase difference; those in (b) are in *antiphase*, the difference being half a wave. A whole period may be defined as involving a phase change of 360°, because a simple sine wave can be modelled as the projection of a steadily moving radius of a circle. So phase differences are sometimes given as angle differences, e.g. 45° in diagram (a) and 180° in (b).

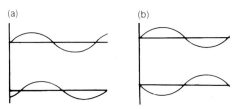

phase *Phase differences of (a) a quarter of a wave and (b) half a wave.*

photocell or photoelectric cell A device that responds electrically to light, either by producing a voltage (*photovoltaic cell*), by emitting electrons (*photoemissive cell*) or by changing its resistance (photoconductive cell). Photocells are used as light sensors in cameras, burglar alarms, automatic doors and lights, etc. *See also* **emission, electron.**

photoelectric effects Effects that occur when the energy of radiation absorbed in matter causes electrical change, with three possible results.

Photoemission is the 'evaporation' of negative electrons from the illuminated surface of a conductor. A certain energy, depending on the surface, is needed to cause an electron to escape. The maximum energy of a photoelectron, therefore, equals the energy of a photon of the radiation minus that escape energy. Photon energy is proportional to frequency; thus, for any surface there is a *threshold frequency* below which there can be no photoemission, however intense the radiation. *See also* **emission**.

Photoconductivity is a property of some **semiconductors**. When illuminated, the material's conductivity rises. (Resistivity falls.) The radiation causes the ionization of atoms in the sample. The extra hole-electron pairs that result allow a higher current. *See also* **conduction, electrical**.

The *photovoltaic effect* appears in semiconductor junctions – the radiation produces a voltage across the junction. This too is a result of ionization. Solutions of electrolytes can show the same effect.

photon   The 'particle' of light and other forms of electromagnetic radiation. It is known that such radiation can deliver energy only in packets, called *quanta* (*singular quantum*). According to the extremely successful **quantum theory**, the photon has both particle and wave properties, and the energy E it carries is linked to its frequency f by the formula $E = hf$, where h is the Planck constant (6.63×10^{-34} Js). In particle physics theory it is the carrier of the electromagnetic force, which affects charged particles.

physical change   A change in a system which one can reverse simply by reversing the energy transfer concerned. An example is the melting of a solid, which can easily be reversed by freezing. *Chemical change* differs in that, although it can also be reversed, that is not a simple task. Thus, while it is not hard to turn a gas into a liquid and back again, it is difficult to reverse the joining of hydrogen and oxygen to make water. Chemical changes always produce different chemical substances (as products) and involve an energy exchange with the surroundings.

physics   The study of matter and energy and how they affect each other. Physics is therefore the widest of all the sciences – it is the basis of chemistry (which in turn is the basis of biology), astronomy, geology and so on, and has mathematics as its main tool.

Until the end of the dark ages of European history, physics was a branch of philosophy, even of religion – people thought about what matter and energy are, but rarely carried out experiments to explore their ideas. (Much earlier, a few Greeks – for instance, Archimedes – had carried out physics experiments).

From the end of the Middle Ages, Galileo, Boyle, Hooke, Huygens and others, especially Isaac Newton, applied mathematics and an experimental approach to the study of matter and energy. However, it was not until the nineteenth century that clear 'modern' concepts of matter and energy appeared. Discoveries were made that could not be explained with the old knowledge. Modern physics started with the discovery of **radio waves**, **radioactivity**, **cathode rays**, and the **photoelectric effects**.

Max Planck (1858–1947) and Albert Einstein are perhaps the greatest figures of modern physics. With many others they created the two main theories of matter and energy: **quantum physics** and **relativity**. Efforts are now being made to produce a single theory of the universe that incorporates these theories.

In practice, physics is usually divided rather artificially into a number of fields. The main fields are **waves**, **energy** (thermal physics), **mechanics**, **electricity** and **magnetism**, **quantum physics**, **relativity**, **particle physics**, **astronomy** and **cosmology**.

pictogram A type of bar chart (*see* **histogram**) whose bars are sets of little pictures that show what the graph plots. Pictograms are quite common in newspapers and primary-school work, but scientists rarely have the need to use them.

piezo-electricity The production of a **potential difference** by applying a mechanical stress to certain crystals. Stress in those crystals causes a voltage between opposite faces. The effect has many uses; they include strain gauges, press-action gas lighters, and microphones. The reverse effect also exists: an applied alternating voltage can make the crystal vibrate; such regular vibrations are used in quartz watches.

pitch The sensation of 'highness' or 'lowness' of a musical note. As a general rule, pitch is related to the fundamental frequency of the sound and the higher this frequency, the higher the pitch. *See also* **quality of sound**.

planet A large solid object in **orbit** round a star. The Earth is a planet in orbit round the Sun as part of the **solar system**. Planets do not produce light but are seen because they reflect sunlight.

The outer solar system contains four *gas giant* planets, Jupiter, Saturn, Uranus and Neptune. These are huge balls of hydrogen, with smaller quantities of helium, methane and ammonia, and with rocky cores. They are circled by numerous moons.

There are four rocky bodies in the inner solar system, called the *terrestrial planets*: Mercury, Venus, the Earth, and Mars. There are also a few satellites, including our **Moon**, thousands of small rocky bodies called *asteroids*, *planetoids* or *minor planets*, and thousands of **comets**.

Beyond the orbit of Neptune orbits Pluto, the small and rocky ninth planet, and its large satellite Charon. Here too circle thousands of small masses of rock and ice, not counted as planets.

To support life, a planet needs liquid water. This may exist beneath the icy surface of Europa, a satellite of the planet Jupiter, and possibly beneath the ice and frozen carbon dioxide at the poles of Earth's neighbour Mars.

To exist on land, life would require an oxygen-containing atmosphere. A small planet does not have enough **gravity** to retain an atmosphere; in addition, planets too close to their star are too hot, while those far away are too cold, so that any 'atmosphere' would lie frozen on the surface.

Planets the size of the gas giants in the solar system have been found circling other stars. As **telescopes** are improved, it will be possible to detect Earth-sized planets, which are believed to exist in large numbers, and to look for evidence of oxygen-containing atmospheres, an important indicator of the presence of life.

See also Appendix E.

plane wave A wave that does not diverge (spread) or converge – the wave crests (where the waves are most intense) are planar (flat).

plasma (in physics) An ionized gas at a high temperature, composed of a mixture of ions and electrons, and considered to be a fourth state of matter, since it does not obey the **gas laws**. Plasmas are formed in **stars**, in **lightning** strokes and in nuclear explosions. Plasmas are used in **fusion** reactors (although such reactors are not yet in practical operation).

plastic behaviour The permanent change in shape or size of some solids when stress is applied beyond a certain value (the *yield point*). It is caused by the slip between layers of atoms when highly stressed; this leads to a form of *flow* in solids. A '*plastic*' is a substance which deforms permanently with ease. It can therefore be shaped using simple equipment. *Compare* **elasticity**.

plate tectonics The processes shaping the Earth's **lithosphere**, which consists of the crust (the planet's outermost layer) and part of the underlying mantle. The lithosphere is made up of a small number of separate plates that slowly move. Where they move apart, molten rock (*magma*) wells up to form the mid-ocean ridges. Where continental plates collide, the crust is folded to form mountain chains. Stresses at the margins of plates cause earthquakes. At a *subduction zone*, one plate is forced beneath another, causing earthquakes and volcanic activity. There was once just one continent, called *Pangaea*. This split into *Laurasia* (giving present-day Europe, North Asia and North America) and *Gondwanaland* (the origin of Africa, India, South Asia, Australasia, Antarctica and South America).

polarization The restriction of the vibrations of transverse **waves** to a single direction only. A transverse wave can vibrate in many directions at right angles to the direction of the wave's movement, as in diagram (a), unlike a longitudinal wave, which can only vibrate backwards and forwards along that direction.

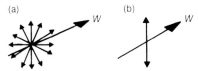

polarization *The vibration of (a) an unpolarized and (b) a polarized transverse wave at right angles to the direction of movement (W).*

When polarized, the vibration of the wave is restricted to one direction only, as in (b). Like all **electromagnetic waves**, light can be polarized; a sheet of special plastic ('Polaroid') will do it, as can **reflection**. Polarized radiations have many uses in science and industry.

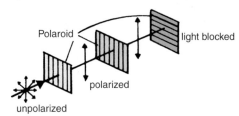

polarization *The polarization of light by Polaroid sheets at right angles to each other; the light is totally absorbed by the second sheet.*

pole 1. A region of a **magnetic field** where the **force** is strongest. Most bar magnets have two poles. A hanging magnet tends to align north-south. Thus the pole that tends to move north is called the north-seeking pole (*N-pole*); the other is the south-seeking pole (*S-pole*).
   2. The geometric centre of a **lens** or **mirror surface**.
   3. The points on Earth or an astronomical body where its axis of rotation reaches the surface.

pollution The effects in the air, in the sea or on the land of poisonous or harmful substances, which make life less safe or pleasant for living organisms. Pollution of water kills fish and other important living creatures; air pollution kills trees and crops and damages buildings. All activities (not just those of people) cause pollution, but it is clearly important to control it. *See also* **environment**, **greenhouse effect**, **ozone layer**.

positron A **fundamental particle** with the same mass as an **electron** but with a positive charge. It is thus the **antiparticle** of the electron.

potential difference (pd) or **voltage** The energy released or taken in per coulomb of charge passing between two points in a circuit. The unit of measurement is the joule per coulomb, given the name *volt*. Note that charge flows as the result of an **electromotive force** (emf) produced by a source of energy such as a battery or a generator, and that emf is also measured in volts. *See also* **Ohm's law.**

potential divider A device or circuit element which uses a pair of resistors or a variable resistor to provide a fraction of the available **voltage** (pd). The pd due to the source is divided across the resistors in direct proportion to their resistances (*see diagrams*).

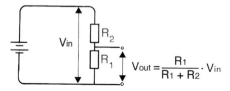

$$V_{out} = \frac{R_1}{R_1 + R_2} \cdot V_{in}$$

potential divider *(a) Fixed resistance potential divider. If $R_1 = 3\Omega$ and $R_2 = 9\Omega$, then $V_{out} = V_{in}/4$.*

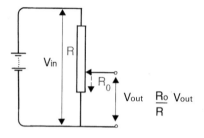

$$\frac{R_0}{R} V_{out}$$

potential divider *(b) Variable resistance potential divider. If output is taken across a quarter of the total resistance R_1 then $V_{out} = V_{in}/4$.*

potential energy The energy possessed by a system in virtue of the configuration of its components, when these act on each other with a force of some kind. Examples of potential energy are:

gravitational potential energy: water at the top of a waterfall, for example, is attracted to the Earth, and can do work and/or gain **kinetic energy** as it falls;

chemical energy: charged particles in a substance gain or lose potential energy (bond energy) in chemical changes, accepting energy or transferring it to the surroundings;

elastic energy: particles in an elastic material attract each other; they therefore gain potential energy when the object is stretched.

power (P) **1.** The rate of energy transfer. Its unit is the watt (W), equal to the transfer of 1 joule per second. One *horse-power* is just under 750 W. Some useful equations for power are:

$$\text{'mechanical' power} = \text{force} \times \text{velocity} = Fv$$
$$\text{dc 'electric' power} = \text{p.d.} \times \text{current} = VI$$

Efficiency is often expressed in terms of power: the useful output power (P_2) divided by the total input (P_1). Thus, in the case of a **transformer**, efficiency can be expressed as:

$$\frac{P_2}{P_1} = \frac{V_2 I_2}{V_1 I_1}$$

2. The measure of how much a lens, mirror or other optical system can focus a beam of rays. High power is the ability to converge or diverge strongly. The unit is the **dioptre**, D. It is given by the equation:

$$P = 1/f$$

where f is the **focal distance** of the lens, mirror or other optical system.

prediction A statement that something should happen in certain circumstances. Predictions can be tested in order to test (and perhaps disprove) any **theory** on which the predictions are based.

pressure (p) The **force** per unit area exerted on a surface. Thus to find pressure, the equation:

$$\text{pressure} = \frac{\text{force}}{\text{area}}$$

is used.

Often, as in diagram (a), the force applied is an object's weight W. The unit is the **pascal** (Pa); 1 Pa = 1 N/m^2.

The pressure at a depth in a liquid, or in a fluid such as the air, equals the weight of the fluid above a unit area, as shown in diagram (b). This gives the relation:

$$\text{pressure} = \text{depth} \times \text{mean density} \times g$$

Note that the pressure at a point in a fluid acts in all directions.

The **gas laws** deal with pressure in gases. *See also* **Archimedes' principle, atmospheric pressure, hydraulic machines, manometer**.

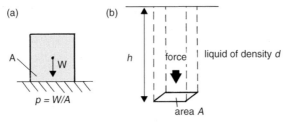

pressure *(a) The pressure of an object on a flat surface; (b) pressure in a liquid.*

pressure cooker A sealed vessel in which food can be cooked rapidly because the water inside boils at a higher temperature than in an open pan. As the **vapour** (steam) cannot escape, it rises in **pressure**; this raises the **boiling temperature**. It is common to obtain a pressure twice that of the air; the water then boils at about 120°C.

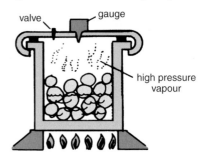

pressure cooker

pressure wave *See* **sound**.

primary colours *See* **colour**.

principal axis *See* **lens, mirror**.

principal focus *See* **lens, mirror**.

prism binoculars An optical device consisting of two simple (astronomical) **telescopes** fitted side by side, one for each eye. The prisms are used as **total internal reflection** mirrors so that light travels backwards and forwards along the telescope tube. The tubes can thus be made conveniently short while still allowing a long light-path and hence good magnifying power. The prisms are arranged so that the final image is upright. The separation of the objective (front) lenses is greater than that of the eyes, thus improving *binocular vision* (three-dimensional vision).

progressive wave *See* **standing waves**.

proof A sequence of steps or statements that establishes the truth of something. It is not possible to prove any theory (or even any **law**) of science. All one can do is show that the **predictions** the theory makes are correct. On the other hand, an experiment can disprove a theory – just by showing that in some given case a prediction based on the theory is not correct. Often in the history of science progress has been based on the testing of predictions, so confirming or disproving theories.

proportional (of two variables) Having a simple relationship. Two variables x and y are directly proportional if, for all values of x, $y = kx$, where k is some constant. This can be written as $y \propto x$, the value of k not being specified. If $y \propto x$, a **graph** of y against x is a straight line passing through the origin (the point $x = 0$, $y = 0$).

  y is *inversely proportional* to x if the graph of y against $1/x$ (or $1/y$ against x) is a straight line through the origin. The equation concerned is $y = k/x$.

  Many laws of physics describe proportionality between two measures. *See*, for example, the **gas laws**, **Hooke's law**, and **Ohm's law**.

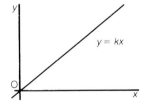

proportional *The graph of y against x when y is directly proportional to x.*

proton A **fundamental particle**, found in the **nucleus** of every atom. The proton has a single positive charge; its mass is one **atomic mass unit**.

  The nucleus of any atom contains one or more protons. The actual number is the **proton number**, Z. In a neutral atom, the number of **electrons** surrounding the nucleus equals Z. *See also* **nucleon number**.

proton number (Z) The number of **protons** in the **nucleus** of an atom. Added to the **neutron number** N, this gives A, the **nucleon number**. Z was formerly known as the *atomic number*. All these measures are pure numbers; they have no units. Any atom can be described in terms of them; their values allow us to predict the chemical and physical properties.

pulley A grooved wheel used as part of a **machine**. In a pulley system an applied force (the **effort**) is carried by a rope via one or more pulleys to move a **load**. A pulley changes the direction of a force, and using several pulleys allows the force to be multiplied when it acts on the load. As with all machines, although the force can be increased the work done by the effort is always greater than the work done on the load. *See* **distance ratio, force ratio**.

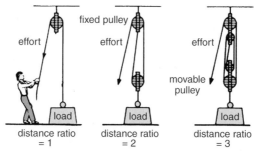

pulley *The addition of pulley wheels increases the force ratio.*

pulsar *See* **star**.

pump A device for moving a fluid, often against gravity. The *lift pump*, shown in diagram (a), can raise water by up to 10m. The *force pump* pushes it to greater heights, (b). Car tyre pumps are rather like the latter, but contain no liquid and are simpler.

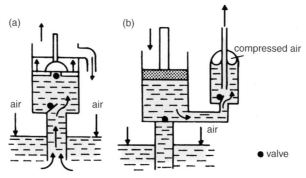

pump *(a) A lift pump; (b) a force pump.*

Q

quality of sound or timbre The characteristic nature of the sound made by a musical instrument. If two different musical instruments play the same note there is a clear distinction between the two sounds. Usually musical instruments give a strong fundamental frequency plus various **overtones**. The proportions of the fundamental and the overtones give the sound its quality. *See also* **sound sources**.

quantum physics A fundamental theory in physics that states that the energy of a system does not have a continuous range of values but can change only by small 'jumps' called *quanta*. The theory was proposed in 1900 by the German physicist Max Planck (1858–1947) to link the internal (heat) energy of a hot body to the pattern of radiation it emitted (*see* **thermal radiation**). In 1905, Albert Einstein (1879–1955) extended the idea to electromagnetic radiation itself, suggesting that the radiation travelled in small energy packets, later called **photons**. He used this idea to explain the **photoelectric effect**, deducing that the photons of radiation of frequency *f* are each of energy

$$E = hf$$

where *h* is **Planck's constant**.

Quantum physics requires that **fundamental particles** sometimes behave like waves, and that **electromagnetic waves** also behave like particles. The quantization of energy is hardly noticeable in everyday life, since the energy systems are so large and quanta are so small; but quantization of energy explains why electrons do not fall into a nucleus, so that atoms are stable.

quark *See* **particle physics**

R

radar A technique for finding the direction and distance of an object by analysing its reflection of **microwaves**. Pulses of **waves** leave the transmitter in different directions: the time before each **echo** returns is measured. Knowing the speed of the radiation and the direction of the echo allows the position of the object to be found.

If the object is moving there is a frequency change during reflection (the **Doppler effect**). That frequency change gives the object's speed, a method used by traffic police when checking motorists' speeds.

radiation Loss of energy from an object, carried by **waves** or **particles** that move outwards from it in all directions. There are many kinds of radiation. For example, *thermal radiation* is energy transfer by **infrared** waves. It is given out by all objects, at a rate that depends on temperature. However, black surfaces radiate (and absorb) these waves better than white ones, unless the temperature is very high.

See also **electromagnetic waves, inverse square law, radioactivity, sound, temperature radiation.**

radio 1. A region of the **spectrum** of **electromagnetic waves**. The lowest defined band is ELF (extremely low frequency), ranging from 30 Hz to 300 Hz, corresponding to wavelengths of $10^7 - 10^6$ m. The highest defined band is EHF (extremely high frequency), 30–300 GHz ($10^{-2} - 10^{-3}$ m). This last band overlaps the **microwave** region.

Band		Wavelength range (m)
LF	low frequency	104–103
MF	medium frequency	103–102
HF	high frequency	102–10
VHF	very high frequency	10–1
UHF	ultra high frequency	1–0.1

The most important use of radio waves is in telecommunications. Radio communication is aided by the *ionosphere* (*see opposite*). This is the part of the upper **atmosphere**, from about 40 km to about 400 km above ground. Many of the air molecules in the ionosphere are ionized, mainly by solar **ultraviolet radiation**. The layer can reflect certain wavelengths radio waves back down to receivers on the ground. *See also* **radio astronomy**.

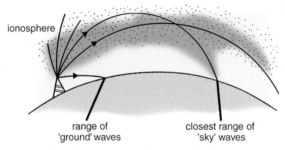

ionosphere

range of
'ground' waves

closest range of
'sky' waves

radio *The reflection of radio waves by the ionosphere.*

2. A device for receiving radio signals.

radio (carrier)
waves induce
signal voltage
in aerial

AERIAL

audio signal

LOUDSPEAKER
changes alternating
current into
soundwaves

TUNER
selects
frequency

DEMODULATOR
rectifies signal and
removes high frequency
radio carrier waves

AMPLIFIER
amplifies
audio signal

radio *Block diagram of the main components of a radio. The incoming signal consists of an audio-frequency signal modulating (shaping) either the amplitude (AM) or the frequency (FM) of a high-frequency carrier wave.*

radioactivity The breakdown (or **decay**) of unstable nuclei to more stable forms. The rate at which this occurs is indicated by the **half-life** period.

Alpha-radioactivity is the release of an **alpha particle** during such a decay. The alpha particle, α, is a helium nucleus. The parent nucleus X changes to a product Y. The mass of Y is four units less than that of X; its **proton number** is two units less:

$$^{A}_{Z}X \rightarrow {}^{(A-4)}_{(Z-2)}Y + {}^{4}_{2}He(\alpha) + energy$$

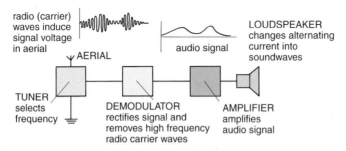

There are a number of forms of **beta decay**, which involves the emission of **electrons**. In the most common, a **neutron** in the parent nucleus changes to a proton; an electron (β-*particle*) escapes. A few radioactive nuclei become more stable by the release of **positrons** (the antiparticles of electrons):

$$^A_ZX \rightarrow {}_{(Z+1)}^{\ A}Y + {}_{-1}^{\ 0}e(\beta) + energy$$

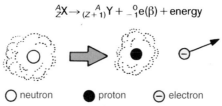

○ neutron   ● proton   ⊖ electron

Gamma (γ) radiation represents a simple loss of energy from the nucleus. A γ-photon appears; it takes with it no charge and very little mass:

$$^A_ZX \rightarrow {}^A_ZX + {}^0_0\gamma + (energy)$$

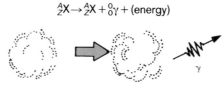

A fourth form of radioactivity is **fission**. In this case the parent nucleus splits into two smaller products with the release of a few neutrons. Here a **chain reaction** may occur.

Note that in some cases the products may be radioactive themselves. It is therefore not easy to provide a pure source of any one form of radioactive radiation. *See also* **atomic radiations**.

radio astronomy The study of stars and other heavenly bodies based on their emissions at **radio** frequencies. Radio waves are produced when **electrons** are decelerated or accelerated, e.g. in the magnetic fields of highly active stars. They are also emitted by molecules such as hydrogen in space. Some astronomical objects do not produce visible light but are strong radio emitters; these include quasars and supernova remnants. The radio waves are collected by radio telescopes, which may be large **dish aerials** or arrays of long parallel wires.

radioisotope An **isotope** that is radioactive. Irradiation of substances with **neutrons** in a nuclear reactor or with **protons**, deuterons or other particles from an accelerator produces radioisotopes. These can be used for the treatment of cancer and as 'tracers' in medicine, and for a variety of tasks in industry. *See also* **radioactivity**.

radiometric dating A method for finding the age of an object which happens to contain a **radioactive** substance. The radioactive substance has a known **half-life**, and by comparing how much of the original element remains in the sample with the new elements produced by its decay, an estimate can be made of the object's age. For example, the radioactive **isotope** of carbon, carbon-14, has a half-life of 5730 years, and is used to date archaeological organic objects; uranium-238 has a half-life of 4.5 billion years and is used to date ancient rocks.

radon A radioactive inert gas produced by the **decay** of various elements. While rare, these elements appear in traces in many rocks. The radon escapes and may reach the ground surface before it decays itself. It is thought that, in some parts of the world, the concentration of radon in buildings is so great that it causes many thousands of cancer deaths each year.

rainbow An effect of **refraction** and **reflection** of sunlight in rain drops. One or more bows, which are actually **spectra** of sunlight, may be seen by someone who has the Sun on one side of them and rain on the other.

rate (in physics) The change in value of some process in unit time (in most cases). Thus the rate of change of an object's velocity (its acceleration, in other words) is the change of velocity in a second; the rate of **decay** of a radioactive sample is the number of nuclei that break down in a second.

Rate is given by the slope of a **graph** of the measure concerned against time.

rate meter A device which measures the **rate** of radioactive decay in terms of the current produced by the emitted particles, either due to the particles themselves or to the flow of ions the particles produce in a gas. Compare this with a Geiger counter, in which individual particles are detected and counted. *See also* **Geiger–Müller tube**.

ray diagrams Diagrams used to locate and describe the **image** of a given object formed by a given optical system. Typical ray diagrams appear in the entries for **lens, microscope** and **mirror**.

In a scale diagram of the system, several rays are drawn from a point on the object, usually the top. Where the rays cross, or appear to cross, is the corresponding point on the image.

When drawing a ray diagram, the following rays behave in simple and exactly known ways:
(a) the ray to mirror/lens parallel to the axis, which is diverted towards the focal point;
(b) the ray passing pass through the focal point, which is reflected or refracted to form a ray travelling parallel to the axis;

(c) the ray to the **pole** passes straight on (for a lens) or is reflected at the same angle on the other side of the axis (for a mirror);

(d) the ray through the centre of curvature of a mirror returns along the same path.

An image is described by stating its position, size, whether it is upright or upside-down, and whether it is real or virtual (whether the rays actually pass through the image or not).

reaction 1. A chemical or physical change that involves two or more inputs and two or more outputs. Chemical reactions involve the atoms or molecules of reagents; these meet and bond together in a new way. The particles in the nuclei of atoms also bond together (in other words, there are forces between them) and are involved in numerous nuclear reactions. In both cases, the reaction may absorb or release energy.

2. An old word for one of the forces in the force-pair described in the third of **Newton's laws of motion**. This law is sometimes quoted as 'Action and reaction are equal and opposite'.

reactivity series The series produced by arranging metals in the order of their chemical reactivity, i.e. how readily they react with other elements. When metals are used as electrodes in electric cells the same order decides the **electromotive force** of the cell: the further apart the metals are in the series the greater the emf.

real and apparent depth An effect produced by refraction to make water or another transparent medium appear to be shallower than it really is. The change of direction due to refraction of rays from an object at the bottom of a pond, say, makes it appear to be nearer the surface.

$$\text{refractive index } n = \frac{\text{real depth}}{\text{apparent depth}}$$

real image *See* **image**.

rectifier A device which passes **current** in only one direction. Most forms are types of **diode**.

The circuit given in the second diagram (*opposite*) is that of a simple rectifier. It is the type used in power packs for toy trains and battery chargers, for example.

The *transformer* input is ac at mains voltage (diagram (a)) at around 230 V.

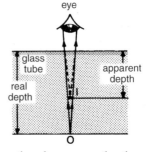

real and apparent depth

The output of the transformer is dc at, say, 12 V. The two diodes form a *full-wave rectifier* with output as shown in (b). A single diode is a *half-wave rectifier*; its output would be as in (c). A *smoothing* **capacitor** and other circuit elements may be used to make the signal as steady as required. Diagram (d) shows a possible output graph from such a circuit.

Direct current is rarely supplied at high power; alternating supply is simpler to transfer via the National Grid. However, there are many uses of electricity which need dc. Examples are battery charging, electroplating and almost all electronic systems. Rectifiers are therefore common.

rectifier *The symbol for a rectifier.*

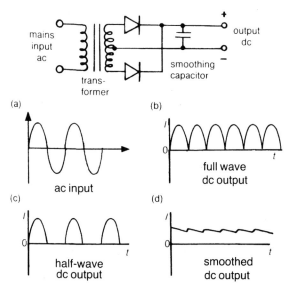

rectifier *Circuit diagram of a simple rectifier with current/time graphs of (a) normal ac input; (b) full-wave dc output; (c) half-wave dc output; (d) smoothed dc output.*

red giant *See* **star, Herzsprung–Russell diagram.**

red shift The lowering of the observed frequency and the associated increase in the wavelength of light due to the motion of observer and light source when they are moving apart. The effect is to move characteristic spectral lines towards the red end of the spectrum or beyond, hence the name *red shift*. Measurement of the shift of spectral lines emitted from astronomical objects allows measurement of their speed relative to Earth, and for very distant objects such as galaxies this can be related to their distances, using **Hubble's law**. *See also* **Doppler effect**.

reed switch A switch consisting of a coil of wire surrounding two 'reeds' – thin strips of easily magnetized nickel–iron sealed in a glass tube. A current in the coil magnetizes the 'reeds', causing them to attract each other, touch and complete a circuit. With an ac supply in the coil the 'make and break' action of the reeds can be as frequent as 1000 Hz. The switch can be used to produce a pulsed current, in telephone systems or as burglar alarm switches, etc.

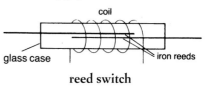

reed switch

reflection The process that occurs when **waves** (such as light or sound) in one **medium** meet the surface of a second and are propagated away from the surface, remaining in the first medium. *See* **absorption, refraction**.

The *laws of reflection* determine where a given *incident* (incoming) *ray* will go after reflection at a given point, the *point of incidence*. Angles are measured from the *normal*, the perpendicular to the surface at the point of incidence, as shown in the diagram. The laws state that:

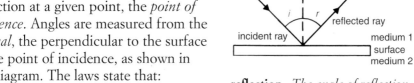

reflection *The angle of reflection, r, equals the angle of incidence, i.*

- the reflected ray is in the same plane as the incident ray and the normal;
- the angle of reflection, r, equals the angle of incidence, i.

See also **diffuse reflection, image, mirror, optical instruments, lateral inversion, total internal reflection**.

refraction The process that occurs when **waves** (such as light or sound) in one **medium** meet the surface of a second and continue on through the second, often in a new direction. *See also* **absorption, reflection**.

The *laws of refraction* specify where a given *incident* (incoming) *ray* will go after refraction at a given point, the *point of incidence*. Angles are measured from the *normal*, the perpendicular to the surface at the point of incidence. The laws state that:

- the refracted ray is in the same plane as the incident ray and the normal;
- for a given pair of media, the sine of the angle of incidence divided by the sine of the angle of refraction is constant (*Snell's law*).

That constant is the **refractive index**. It relates to the speeds of the radiation in the two media.

See also **colour, lens, mirage, optical instruments, rainbow, total internal reflection**.

refractive index or refractive constant (*n*) A dimensionless number (that is, one having no units) which determines to what extent a ray of light is refracted (changes direction) on going from one medium to another. **Refraction** is due to a difference in the speeds of light in the two media, and the refractive index *n* is linked to both speeds and to the change in direction by the formulae:

$$n = \frac{\text{speed of light in medium 1}}{\text{speed of light in medium 2}} = \frac{\sin(\text{angle 1})}{\sin(\text{angle 2})}$$

For light going from air to glass, the formula becomes:

$$n = \frac{\sin i}{\sin r}$$

as illustrated in diagram (a). Diagram (b) shows how a change in speed of waves produces a change in their direction. This may be seen using a **ripple tank**.

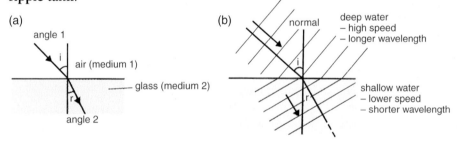

refractive index *Refraction of water waves (b) illustrates a similar effect with light (a).*

refrigeration A technique for **energy** transfer from a cool box to the warmer outside air. As this is not the normal direction of net energy transfer (heat normally flows from a warmer region to a cooler one), a motor is needed to supply extra energy (*see diagram*). This pumps a suitable vapour round a closed loop. The pump raises the pressure; the vapour becomes liquid and gives off **latent heat**. At this

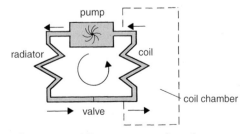

refrigeration *The operation of a refrigerator.*

stage the liquid goes through the black radiator coil outside the cool chamber. It is forced through the valve to the low pressure side of the loop. This makes it evaporate, absorbing heat from the cool chamber to do so.

regelation The refreezing of ice after melting, especially when the melting is caused by an increase of pressure that is subsequently removed. This is how ice skates work – the pressure of the thin blade melts the ice, which then acts as a lubricant, refreezing when the skate moves on. A ball of snow, consisting of tiny ice crystals, partly melts when squeezed. Regelation when the pressure is removed makes the ball hold together.

relative density *See* **density**.

relative permittivity *See* **dielectric**.

relativity Two theories that describe matter, space and time and how they relate to each other. They were developed by Albert Einstein (1879–1955) between 1905 and 1915. The special theory (published in 1905) states that the laws of physics are the same for all people moving at constant velocity relative to each other. In particular they would all find the same value of the **speed of light** c (about 3×10^8 m/s).

The theory predicts, among other things, that the mass and length of a moving object, and the rate of processes associated with it, depend on its observed speed. Also, no mass can be forced to travel at or faster than c. The well-known equation $E = mc^2$ follows too. The latter means, in effect, that mass and **energy** are equivalent. No test has given cause to doubt these concepts; indeed they are part of daily experience in many fields of science and technology. *See also* **mass–energy**.

The *general theory of relativity* (1915) is mainly concerned with **gravity**. It predicts that light curves near a massive object and that its wavelength also changes in a gravity field. Again, tests bear these statements out.

Today the greatest challenge facing physicists is to unite relativity with **quantum physics**.

relay A device based on an **electromagnet** which allows a small **current** to control a large one. When a small current is passed through the electromagnet (*see diagram*), the core attracts the plate. As this moves, it closes the main circuit switch. Cutting the small current opens the main circuit.

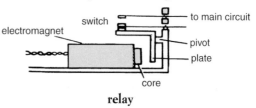

relay

reliability The extent to which the outcome of an experiment can be trusted or not. The reliability of an experiment in science depends on its design and on the analysis of the results. It needs great care to ensure that an experiment explores exactly what is wanted and obtains accurate and relevant data. *See also* **valid data**.

renewable energy source *See* **alternative energy**.

resistance (R) A measure of the opposition of a circuit element or section to the flow of charge (**current**, I). The unit is the *ohm*, Ω: $1\ \Omega$ is the resistance of a circuit element passing 1 A when the voltage V between the ends is 1 V. R can be measured, using the relation $R = V/I$, with an ohm-meter, a potentiometer, or a circuit like that shown in the diagram.

All normal circuit components have resistance. As they oppose current, work is done and energy is converted into thermal form (*see* **fuse**). The **power**, P, involved is the rate of heat generation; it is given by VI, V^2/R or I^2R.

Temperature often affects a sample's resistance. In the case of most metals, temperature rise makes R rise; with **semiconductors** and insulators R falls.

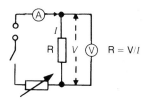

resistance *A circuit to measure resistance.*

resonance Large-amplitude vibration of a system when driven with a frequency at or close to its **natural frequency**. If someone sings a note near a piano, some strings will resonate, or vibrate 'in sympathy'. In the same kind of way, the *tuning circuit* of a radio set will pass high-amplitude signals only if their frequency is near the circuit's natural value. Resonance is often the cause of the collapse of buildings in earthquakes. These examples of resonance match others in the fields of electronics, radiation and nuclear physics, for example. In a system resonating at its natural frequency, **standing waves** may be set up. *See also* **sound sources**.

resultant The single vector that produces the same effect as a combination of two or more vectors. For example, even though more than one **force** may act on an object at rest, the object cannot move in more than one way. If it moves, the motion will be as if only one force acted. This net force is the resultant of the actual forces. The *equilibrant* of the set of forces is defined as equal and opposite to the resultant.

Several special cases should be noted (*see diagrams at* **equilibrium**):
(a) The forces act along one line. The resultant is their algebraic sum. The resultant acts along the same line.

(b) The forces are parallel. The resultant is their sum. The resultant acts along a line given by the law of **moments** (torques).
(c) The forces act at an angle. The resultant follows the **parallelogram rule** of vectors.
See also **component**.

retardation *See* **acceleration**.

Richter scale *See* **earthquake**.

right-hand rules Two rules specifying the direction of magnetic effects. The *right-hand grip rule*, shown in diagram (a), gives the field direction (shown by the fingers) round a current (thumb). *Fleming's right-hand dynamo rule*, (b), shows the directions of **f**ield (**f**irst finger), **m**otion (thu**m**b), and **c**urrent/voltage (se**c**ond finger) in cases of **electromagnetic induction**. Compare the latter with the **left-hand (motor) rule**, used to give the direction of the **motor effect**.

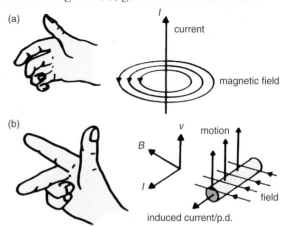

right-hand rules *(a) The right-hand grip rule; (b) Fleming's right-hand dynamo rule.*

ring main or ring circuit A modern house circuit where the cable loops from the **fuse** box round to a series of 13A outlets and back to the fuse box. *See diagram.* The main fuse limits the current to 30 A.

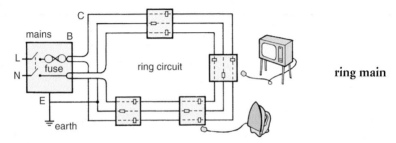

ring main

ripple tank A device which shows how water waves behave. The source is a vibrating object dipping into the surface. The edges of the tank absorb the waves to prevent confusing reflections. A lamp below the glass base projects an image of the waves onto a screen.

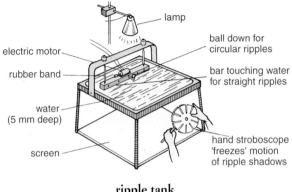

electric motor
rubber band
water (5 mm deep)
screen
lamp
ball down for circular ripples
bar touching water for straight ripples
hand stroboscope 'freezes' motion of ripple shadows

ripple tank

rock The solid mass of mixed minerals that make up the Earth's crust. Rocks are of three main types:

(a) *igneous*: those like granite or basalt that have risen from below the crust and have solidified from a hot, liquid *magma*;

(b) *sedimentary*: rocks in layers laid down by wind (e.g. loess), or on the sea floor (e.g. sandstone, formed from eroded rock fragments; limestone, formed by chemical deposition from sea water; or chalk, formed from the skeletons of minute living things);

(c) *metamorphic*: rocks hat originated by processes (a) or (b) and were then changed in nature as a result of pressure and/or high temperature.

Any kind of rock may be weathered (broken into small pieces) and eroded (the fragments carried away by e.g. wind or water – *see* **erosion**, **weathering**. These are laid down to form sedimentary layers of, for example, sand and gravel. New layers on top compress these and turn them into thin, hard rock layers deeper and deeper in the crust. Thus mud may turn into shale, or even into the metamorphic form known as slate. Movements in the crust may bring these deep layers back to the surface; and the weathering and erosion starts again. The ocean floor, with its accumulated sediments at the edge of one tectonic plate, may sink beneath overriding continental crust on another plate. The rock may melt at depth and form a magma, which may reach the surface again through volcanic activity, thus completing the *rock cycle*. *See also* **plate tectonics**.

rocket engine An engine that propels a vehicle by emitting hot gases at high speed. Rockets are a simple example of the third of **Newton's laws of motion**, in that the force expelling the hot gases is always accompanied by an equally strong force acting on the vehicle, pushing it in the opposite direction. The **momentum** given to the gases is equal and opposite to the momentum gained by the vehicle. Rocket engines carry both fuel and the oxygen to burn it, in the form of a chemical called an oxidizer, and so they can be used equally well in the vacuum of space and in the atmosphere.

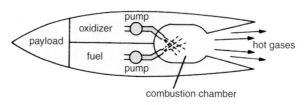

rocket engine

Rutherford-Bohr model The model of the atom as a massive positive nucleus surrounded by a cloud of electrons, proposed by Ernest Rutherford (1871–1937) in 1911. This model was extended by Niels Bohr (1885–1962) using the **quantum theory** to explain how the electrons could orbit the nucleus without spiralling inwards.

S

salt The product of the reaction of an acid and an alkali, a compound in whose simplest form the molecules consist of two oppositely charged **ions**. Most minerals are salts, a well known example being NaCl, sodium chloride, which is the common salt found in many foods. When molten or in solution, salts conduct electricity well.

Sankey diagram A diagram used to illustrate the flow of energy through a system, identifying the various systems to which energy is successively transferred.

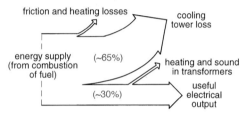

Sankey diagram *Energy flow in a typical power station.*

satellite An object in **orbit** in the gravitational field of a second, more massive object. Thus the Earth is a satellite of the Sun (but, as the Sun is a star, the Earth is called a **planet**); the **Moon** is a satellite of the Earth. The planets of the solar system have over 60 known satellites. The Earth has thousands of artificial satellites, put into orbit by rocket over the past few decades. These aid research and communication, survey and map the surface, or are weapons.

saturation The state of a system when some process is complete. Thus in a saturated solution, the solvent can hold no more solute. Saturated air can hold no more water **vapour**. A saturated magnetic substance cannot be magnetized more strongly (*see* **domain**).

scalar Describing a quantity with size but no direction. Examples are mass, volume, density, energy, and temperature. Compare **vector**, a quantity that has both size and direction.

scaler An electronic device that represents the output of a radiation detector as a rate of charge or of energy transfer. A scaler can be used rather than a **counter** with a **Geiger–Müller tube**, to measure the ionizing radiation from a radioactive substance.

seasons Divisions of the year in parts of the world distant from the equator, marked by different lengths of day and night, different quantities of solar radiation and (as a result) different wind and weather patterns. The effect is a result of the fact that the Earth's axis of spin is tilted by 23° to its plane of orbit. In the winter this makes the days short and the nights long, and the angle of tilt also helps to reduce the input of radiation energy from the Sun; in the summer, the reverse is the case.

secondary emission The release of **electrons** from a surface caused by absorption of energy from a beam of particles (such as other electrons). *See also* **emission, electron, absorption**.

seismic waves The shock waves produced by an **earthquake**. There are two main types of seismic wave: *surface waves* and *body waves*. Surface waves are slow oscillations of the upper crust of the Earth. Body waves travel through the Earth, and are in turn of two kinds:
(i) *P-waves*, which move through rocks as longitudinal (push-pull) waves;
(ii) *S-waves*, which move through rocks as transverse (sideways) waves;
   S-waves cannot travel through a liquid, so the fact that they are not observed on the far side of the Earth from an earthquake is evidence that the Earth's (outer) core is liquid. *See also* **Earth, earthquake**.

seismograph or **seismometer** Instruments used to detect and measure **seismic waves**.

semiconductor A substance able to conduct charge to some extent – less than a metal but more than an insulator. *See* **conduction, electric**. The resistance of a semiconducting sample falls with temperature rise; it also falls if it contains certain impurities.

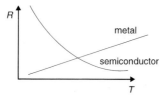

semiconductor *A resistance/ temperature graph comparing metals with semiconductors.*

   The most common semiconductors are silicon and germanium. A few of the atoms in their crystals lose electrons; these then wander through the crystal. A small rise in temperature releases more electrons, so that resistance decreases. Adding small amounts of the correct type of impurity provides more *charge carriers*. An *n-type* semiconductor has many negative free electrons; a *p-type* substance conducts charge by positive 'holes'. (A **hole** can be thought of as a missing electron.)

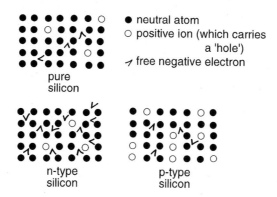

neutral atom
○ positive ion (which carries a 'hole')
↗ free negative electron

pure
silicon

n-type
silicon

p-type
silicon

semiconductor   *Types of semiconductor material.*

The semiconductor **diode** is a *p–n junction*; a piece of p-type material is in close contact with a region of n-type. When the junction is first established, a flow of electrons and holes across the junction takes place for a short time, as electrons combine with holes. This sets up a voltage which thereafter allows current one way only.

A **transistor** can be thought of as two p–n junctions back to back. *See also* **integrated circuit**.

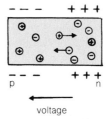

semiconductor
A p-n junction diode.

sensor   A device that takes in and reacts to some kind of input energy, and outputs a related electrical signal. As input energy can arrive in many different ways, there are very many kinds of sensor, and often very many subtypes of each. Examples are sensors for acidity, light level, magnetic field strength and temperature. Sensors are crucial in such cases as the **control** of a process in a factory, surveillance, and robots. They are used widely to collect measurements in **data logging**. Examples include the **microphone, thermocouple, thermistor, light-dependent resistor, potential divider** (voltage divider).

series   In electrical technology, describing components joined in a circuit so that the same **current** flows in each. The **voltage** between the ends of the group is the sum of those between the ends of each element. The resistance R of the group of resistors in diagram (a) is the sum of their resistances:

$$R = R_1 + R_2 + R_3.$$

In the case of capacitors in series, as in (b),

$$1/C = 1/C_1 + 1/C_2 + 1/C_3.$$

Compare **parallel**.

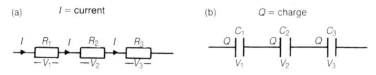

series *Resistors (a) and capacitors (b) in series.*

shadow A dark space behind a solid object from which light is blocked. As light travels in a straight line in a given **medium**, an opaque object will cast a shadow: radiation cannot travel behind it. However, see **diffraction** for exceptions to this general rule.

   If the source is larger than a point, the shadow is not sharp. The central *umbra* receives no rays; the outer *penumbra* is reached by some radiation. *See also* **eclipse**, **transparent**.

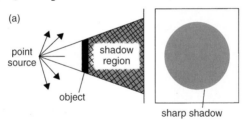

shadow *(a) A point light source casts a sharp umbra; there is no penumbra.*

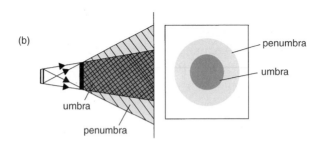

shadow *(b) Shadow produced by a large source of light.*

shear force A force that causes a distortion of shape. For example, spreading a pack of playing cards over a table-top involves applying shear forces.

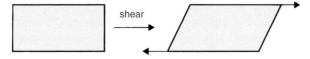

shear force *The change in shape caused by shear forces.*

shock, electric A set of symptoms that follow the passage of **current** through the body. They may include feelings such as tingling and burning, muscle spasms, skin damage, unconsciousness and death.

short circuit A low-resistance path between two points in a **circuit**. The large flow of charge resulting draws a heavy current from the source. If the short is a fault, the **fuse** should break the circuit.

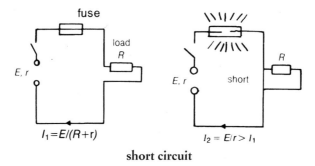

short circuit

short sight *See* **vision defects.**

signal A quantity that varies with time to allow the transfer of information. Signals can involve changing **voltage**, changing **current**, or waves changing in frequency or amplitude. **Electronics** can be regarded as the study of how circuits and circuit elements affect input signals.

signal generator An electronic device which produces an alternating voltage output over a wide range of frequencies. Audio signal generators produce frequencies in the range from a few hertz to about 50 kHz, radio-frequency generators work at much higher frequencies (up to several MHz).

significant figures The number of digits quoted in a quantity or measure, which should match the accuracy to which the value is known. If a length is measured to the nearest millimetre, its value should be quoted as 1.245 m, for example – that is, to four significant figures. When numbers are combined mathematically, the final result should be quoted to the same number of significant figures as the least precisely known number. For example, a calculation of speed, 1.245 m ÷ 2.1 s, might yield 0.592857142 on a calculator. This should be rounded off and written 0.59 m/s (i.e. to two significant figures, as the value for time is given to only two significant figures).

SI units *See Appendix A.*

slide projector A device for displaying, on a screen, images of transparent slides. It works by focusing a real inverted **image** of the transparent slide on a distant screen. In the diagram the mirror, (a), and the *condenser lens*, (c), concentrate light from the source, (b), onto the slide, (d). This is the **object** whose light is refracted by the *projection lens*, (e); the image appears on the screen and is focused by moving the projection lens.

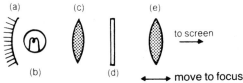

slide projector *The components of a projector; (a) mirror; (b) bulb; (c) condenser lens; (d) slide; (e) projection lens.*

Strip and movie projectors work on the same principle as do overhead and opaque projectors. In the latter, light is reflected from an opaque object rather than passing through a transparent one.

solar energy Energy from the Sun, coming mainly in the form of electromagnetic radiation (such as light and **infrared** and **ultraviolet** radiations). Less than one billionth of the Sun's total output reaches the Earth and the atmosphere reflects 37% of that. On average, the ground receives about 800 W per square metre. Plants use that (at less than 0.5% **efficiency**) and there are various ancient human uses (such as drying leather and leaves and getting salt from sea water by **evaporation**).

Modern methods of using solar power involve more efficient techniques of absorbing the energy in water or **photocells**. Some of these systems approach 35% efficiency. Wind- and water-power systems also involve solar energy, which drives the wind systems and the **water cycle** on Earth, as well as the main global ocean currents.

solar system The Sun (a small normal star near the edge of our **galaxy**) and the 'family' of **planets**, comets, smaller chunks of rock, and gas and dust clouds held by its gravity. The scale of the system is often measured in *astronomical units* (AU). 1 AU is 150 million kilometres, the distance between the Sun and the Earth. On that scale, the Sun's disc is 0.01 AU across; the orbit of Pluto, the furthest known planet, has a radius of 40 AU. There are many comets and other bodies far more distant than that; the system itself is about 100 AU in radius. The next nearest star is some 250,000 AU away.

See Appendix E for data about the solar system.

solenoid A cylindrical coil of wire whose radius is smaller than its length. When a current is passed through the coil a **magnetic field** is produced inside and around it; it can thus be used as an **electromagnet**.

solution A uniform mixture of the particles of a solid (*solute*) with those of a liquid (*solvent*). The solvent may also be a second solid, as with an alloy. The process of solution may take place in different ways: often the solute forms **ions** in solution. In most cases there are energy changes involved.

sonar A system very like **radar**, but which uses ultrasonic waves travelling through water rather than radio waves travelling through the air or space. (Radio waves do not travel far in water.) Sonar is used to detect submarines, underwater obstacles or shoals of fish. It is also used to find the depth of the water using reflections from the bottom. *See also* **echo**.

sound A form of radiation which involves *pressure waves* – regions of alternating high and low pressure travelling through some material. The vibrating source pushes particles in the material closer together; they then move further apart, so producing a series of *compressions* and *rarefactions* – a sound **wave**.

For a typical musical sound, a plot of pressure in the medium at different points at a given time is a sine wave, or a combination of different sine waves. So too is a plot of the pressure at a point as time passes.

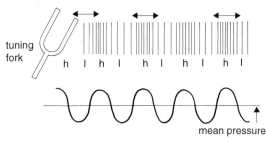

h - high pressure (compression)
l - low pressure (or rarefaction)

sound *Plots of a typical sound wave.*

Sound shows most wave properties: **absorption, reflection, refraction, diffraction** and **interference**. However, sound is a longitudinal wave, involving backward-and–forward vibrations, and so cannot show **polarization**. Pressure waves, like all others, have a frequency **spectrum**. The range which the human **ear** perceives as sound is from about 20 Hz to 20,000 Hz; *See also* **infrasound, ultrasonics, sound sources, speed of sound**.

sound level *See* **decibel**.

sound sources Objects that vibrate in a medium to produce sound waves. The **frequency** of the sound produced by a source depends on its nature and how it vibrates. In the case of some sources, such as bells, the sound wave emitted can be very complex.

The source's simplest mode (essentially, its frequency) of vibration is called its *fundamental*. In the case of a taut string or wire, the fundamental is as in diagram (a). The ends of the wire are fixed and the centre moves most; **nodes** (N) and **antinode** (A) are found at the ends and in the centre respectively. The wavelength, λ_0, of the sound radiated is twice the length l of the wire. *See also* **standing waves**.

Another important sound source is the *closed pipe*. This is a tube closed at one end, like a glass bottle or an organ pipe. The fundamental mode of vibration of the air column inside has a node at the closed end (air is prevented from vibrating here) and an antinode at the open end. Now λ_0 is four times the tube length l. The distance between the dashed lines in diagram (b) gives an idea of the variation of the amplitude of vibration of the particles of air at each point; however, the actual amount of vibration is tiny and is in the forward-backward direction.

The diagrams show the fundamental modes of vibration of two sound sources. The wavelengths radiated, λ_0, relate to each **natural frequency**, f_0, by the wave equation $c = f_0 \lambda_0$; c is the **speed of sound** in air.

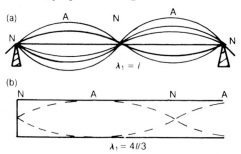

sound sources *Nodes and antinodes in the first overtone.*

However, more complex modes of vibration can also occur. These produce **overtones**. The *first overtones* for a taut string and an open pipe are shown in the second pair of diagrams. Note how the wavelength λ_1, relates to source length in each case.

The following pair of diagrams shows the second overtone for each source.

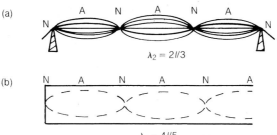

sound sources *Nodes and antinodes in the second overtone.*

Few traditional sound sources produce pure notes of a single frequency only. A good **signal** generator (electronic oscillator) driving a good speaker can do so. The **quality** (or *timbre*) of the sound from a source depends on the mix of overtones and fundamental. *See also* **music**, **noise**, **pitch**.

source resistance (r) The resistance of a source of electrical energy. *See* **internal resistance**.

space travel Travel beyond the Earth's atmosphere. The first complete orbit of the Earth was accomplished in 1957 with the unmanned Russian satellite *Sputnik*. The first manned space flight, carrying a Russian, Yuri Gargarin, was in 1961. The American Neil Armstrong set foot on the Moon in 1969. Most spacecraft do not carry people, although a long-term aim is for people to travel to other parts of the **solar system** to live. The American *Pioneer* and *Voyager* unmanned spacecraft travelled outside the solar system after making close flybys of the outer planets, from which they sent photographs and valuable scientific data.

The chemical rockets that are used at the moment for space travel are very inefficient and costly. Other techniques may allow easier space travel in the future.

spark An electrical **discharge** through a gas that lasts for a very short time; it is accompanied by a flash of light and a sharp crackling noise. A channel of gas **ions** carries the current. The energy transfer leads to chemical change, temperature rise (giving the sound), and electromagnetic waves. *See also* **lightning**.

spark counter A **radiation** detector able to detect **ionizing** particles. **Sparks** appear across a 5 kV gap where ions from the source pass through, and the sparks can be counted by an electronic circuit.

specific (in physics) 'Per unit mass'. Thus the **specific thermal capacity** of a substance is the thermal capacity of unit mass (1 kg). *See following entries.*

specific charge The **charge/mass** ratio, Q/m, of a **particle**. The unit is the coulomb per kilogram. To measure this helps the researcher to identify the particle. For an electron Q is 1.6×10^{-19} C; m is 9.1×10^{-31} kg. The specific charge is therefore 1.8×10^{11} C/kg.

specific thermal capacity or specific heat capacity (c) The energy involved in changing the temperature of a kilogram of a substance by 1°C. The unit is the joule per kilogram per degree Celsius, J/kg°C or joule per kilogram per kelvin, J/kgK. The values for some common substances are given in *Appendix B*.

In a given case, c can be found with the **method of mixtures** or by using a **joule meter** to measure energy supplied, E. Then $c = E/m\Delta T$, where m is the sample mass and ΔT is its temperature change.

See also **thermal capacity.**

specific latent heat (L) The energy involved in the **state change** of a kilogram of substance. *See* **latent heat.** The unit is the joule per kilogram, J/kg. The energy E involved in the state change of a sample of mass m is $E = mL$. The values of L for water are 336 kJ/kg (ice/liquid) and 2270 kJ/kg (liquid/vapour). L can be found either with the **method of mixtures** or by using a **joule meter** to measure E.

spectrum The band of colour produced by spreading out visible light according to its frequency or wavelength; or a similar distribution of the whole range of electromagnetic radiation in general; or an analysis of any wave motion, set of particles, etc. according to energy or some other physical characteristic. The **rainbow** is a spectrum of sunlight; a *voiceprint*, which is a graph of the energy of the frequencies present in a sample of speech, is a spectrum of the sound energy.

In the case of electromagnetic waves, an *emission spectrum* shows the radiations from an actual source. An *absorption spectrum* shows the radiations absorbed by matter between source and detector. Thus the Sun gives a continuous emission spectrum; gases in its atmosphere and the atmosphere of the Earth absorb some wavelengths to give an absorption spectrum.

The light spectra of simple gases consist of series of lines, *line spectra.* Molecules give *band spectra* (blocks of close lines); hot dense matter gives a *continuous spectrum* of all wavelengths.

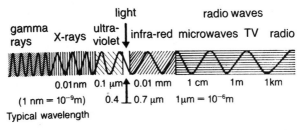

spectrum *The electromagnetic spectrum.*

speed (u, v) The rate of change of distance with time. The unit is the metre per second, m/s. Speed is a **scalar** measure as its value is independent of direction. **Velocity**, on the other hand, is a **vector**, being the rate of change of distance with time in a given direction.

speed *Travelling between A and B, speed is calculated by distance (s)/time (t).*

speed of light The speed at which electromagnetic **radiation** travels in a **vacuum** (such as outer space). The speed of light in a vacuum is now *defined* to be 2.99792458×10^8 m/s, which is taken to be 3.0×10^8 m/s in elementary calculations. This value is used to define the basic standard of distance, so that a metre is the distance travelled by light in $1/c$ seconds (1/3,335640952 s). **Relativity** predicts that no object can accelerate to the speed of light in empty space. Light travels more slowly when it enters a physical **medium**; the speed depends on the substance and the wavelength of the light. *See also* **refractive index**.

speed of sound The speed of sound and other pressure waves through matter. This depends on the **elasticity** of the medium and its **density**. High density lowers the speed, high elasticity increases it. Although the density of solids and liquids tends to be far higher than that of gases, the elasticity is higher still. As a result, the speed of sound is lowest in gases. For example, the speed of sound in air is 300 m/s while in water it is 1500 m/s and in steel is 6000 m/s.

The speed of sound in a given substance rises with **temperature**. The speed of sound can be measured either by timing the reflection of a short pulse of sound or by setting up **standing waves** in a sample.

Objects moving through air at a speed greater than that of sound have *supersonic speeds*. Shock waves form around them; these cause the *sonic boom* of a supersonic aircraft and the crack of a long whip.

stability   A system's tendency to stay in **equilibrium**. A system is stable if a disturbance sets up a reaction which restores the system to its first state. For example, consider a ball rolling on a surface, as shown in the diagram. The restoring force is **gravity**. A disturbance in (a) raises the object's **centre of mass**, (b) lowers it, or (c) does not change its level. The object will tend to (a) return to its first state – stable equilibrium; (b) move away from it – unstable equilibrium; (c) stay in any new position to which it is moved – neutral equilibrium.

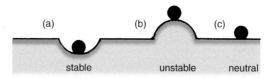

stability   *Stable equilibrium (a); unstable equilibrium (b); and neutral equilibrium (c).*

standard form or scientific notation   The expression of numbers in a form such as 1.25×10^2, where the decimal point follows the first digit. Thus numbers in standard form would be from 1.00×10^n to 9.99×10^n. This way of expressing numbers is best when dealing with very large numbers like the speed of light, 3×10^8 m/s, or very small numbers, like the charge on an electron, 1.6×10^{-19} C.

standing waves or stationary waves   **Waves** that result from **interference** between two similar waves of equal wavelength moving in opposite directions through the same space. Standing waves often appear when a *progressive* (or travelling) *wave* is reflected back along its path.

Different points on a progressive wave have the same amplitude but differ in **phase**. The reverse is true in the case of standing waves.

All **electromagnetic waves** can set up standing waves, as can those on the surface of water. They are also involved in the action of all **sound sources**. Thus, when one plucks a taut wire, it vibrates. Waves pass to and fro between the ends of the wire and give rise to a standing wave. *See* **nodes**.

star A large, luminous globe of gas (mostly **plasma**) at a high temperature, which emits radiation. Stars gain their energy by converting particles of matter into both kinetic energy and radiation. Most of the stars we see in the sky convert hydrogen into helium by the **fusion** process. Stars do not last for ever. As time goes on the supply of available hydrogen is used up and other reactions may then take place, depending on the mass of the star. The more massive the star, the higher its internal temperature and pressure and the faster it uses up its hydrogen. As long as it is in the so-called 'hydrogen-burning' stage it is likely to be stable and to keep a constant temperature and rate of emission of radiation. This means that it stays in one place on the main sequence of the **Hertzsprung–Russell diagram**.

A star begins as a large cloud of gas and dust. Gas clouds (and so stars) are mostly hydrogen and helium in the ratio of about 3 to 1. As time goes on the gravitational self-attraction of the cloud makes it contract. As the cloud falls in on itself it gets denser and also hotter, as the original gravitational potential energy becomes kinetic energy of the particles. When the centre of the protostar is dense enough and hot enough (at 10 million K), nuclear fusion begins. Immediately, the collapse stops as the pressure of the core gases is kept high enough to withstand the gravitational force. Energy from the nuclear reaction eventually rises to the surface (via convection and radiation processes) and the star becomes very luminous. The *luminosity* of the star is the total radiation energy it emits, and whilst the star is stable equals the energy released in the core from nuclear fusion. There is very little direct evidence about what stars are like inside, or about *stellar evolution* – how stars change with time. The following theory relies on computer models and a knowledge of nuclear reactions.

A star like the Sun will have enough hydrogen to keep stable for about 10 billion years. The Sun is now about 5 billion years old. When hydrogen is no longer available, the temperature and pressure in the core decrease and the star begins to collapse. The collapse heats up the core again and brings fresh hydrogen into play. This reacts in a shell around the collapsing core and heats up the outer layers so much that they expand to many times the original size of the star. For example, if this happened to the Sun it would expand to beyond the Earth's orbit. The result is a larger, cooler star, no longer white-hot but red-hot – a *red giant* star.

An ordinary red giant star will survive by the fusion of helium and other nuclei into heavier nuclei, then gradually cool and contract (possibly with an explosion every few thousand years) and its core will become a very dense, small, hot remnant – a *white dwarf* star.

Stars more massive than the Sun will go a stage or two further in the nuclear fusion process until the core is mostly carbon. The fusion of carbon into larger nuclei is a very fast process and when the required temperature is reached the result is a huge explosion, in which tremendous energy is generated and the outer layers of the star are blasted into space. We see the result as the sudden appearance of a very bright new star – a *supernova*. The core is left as a very hot and incredibly dense object composed entirely of neutrons – a *neutron star* – more massive than the Sun but only 20 km in diameter. Rotating neutron stars emit beams of radio waves which are detected on Earth as pulses – such stars are called *pulsars*. The Crab Nebula is a pulsar, and the remains of its exploded outer layers can be seen in telescopes. It was seen as a supernova by Chinese astronomers in 1054.

state The physical structure, form or condition of a substance. A substance can exist in the solid, liquid, **gas** or **plasma** state. These differ greatly in the way they behave. The **kinetic model** describes the states as different arrangements of particles. The pattern at a given moment depends on the substance and on the temperature and pressure of the sample.

In solids, the particles hold each other in a fairly rigid pattern as shown in diagram (a). They have little energy: the temperature is low. They vibrate about their mean positions. As the temperature is low, so is the **evaporation** rate.

Diagram (b) shows the particles in a liquid. They are not fixed in place but move around; thus a liquid can flow to take the shape of its container. However, the forces between the particles are still strong enough to allow only a few to escape the surface. There is also a **surface tension** but this still allows more evaporation than in solids.

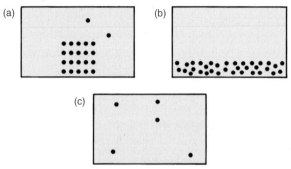

state *The arrangement of particles in a solid (a), a liquid (b) and a gas (c).*

The particles of a gas or a **vapour**, shown in diagram (c), are almost fully free. They move at high speed through the whole space available. *See also* **state change**.

state change The change of matter from one physical state (solid liquid or gas) into another. The change is always accompanied by the evolution or absorption of energy.

For a given pure substance at a given pressure, a change of state always occurs at the same temperature (this is not so for **evaporation**) Impurities and outside pressure both affect state change temperatures. Thus a liquid boils at a higher temperature if there is something dissolved in it, and if the pressure is higher than normal. (*See* **pressure cooker**.)

In most temperature/pressure situations, a substance can exist in only one state. The graph (*see below*) shows the conditions at which state change can occur. *See also* **vapour**.

P is the *triple point*, the only temperature/pressure combination at which solid, liquid and vapour can exist in equilibrium. The triple point of water is at 0.01°C and 611 Pa.

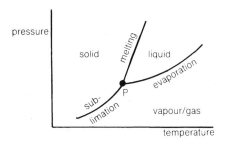

state change *A pressure/temperature graph showing the triple point (P) at which a solid, liquid and vapour can exist in equilibrium.*

static electricity Stationary electric **charges**, as opposed to moving ones, i.e. **current electricity**. 'Static' often appears when an insulator is rubbed; the charge may then remain a long time. The effect causes problems with modern furnishings, in mines and factories, and in aircraft fuelling systems.

statics The study of objects and systems in **equilibrium** under the action of outside **forces**.

stellar evolution *See* **star**.

stethoscope A medical device used to listen to the sounds produced in the interior of the human body, such as in the heart and lungs.

strength A measure of how well a sample of a substance can resist being *deformed* (such as being stretched or compressed) or fractured.

Sun A typical (although relatively small) **star** near the edge of our **galaxy**. It is a hot ball of gas, with a temperature of 20 million K at the centre and 6000 K at the visible surface (although there is a low-pressure atmosphere that extends much further out). Nuclear reactions in the Sun produce approximately 4×10^{20}MW (a big power station may output 10^3MW). *See* **solar energy**, **solar system**, **star**.

The Sun moves round the centre of the Galaxy (our own galaxy is given a capital letter to distinguish it from others), and with the Galaxy through space. The Earth moves round the Sun (taking one year, *see* **seasons**). The Earth also spins on its axis once a day, which makes the Sun 'rise' over the Eastern horizon in the morning and go behind the Western horizon ('set') in the evening. All the other astronomical objects in the sky appear to circle the Earth once a day in the same way.

superposition, principle of *See* **interference**.

superconductivity The state of zero resistance found in many substances at very low temperatures. A current once started in a superconducting loop will thus go round and round for ever (as long as the low temperature remains). No expenditure of energy is needed to maintain the charge flow and no energy is released by it. Extensive research is being carried out to use this effect to carry electrical power and currents in computers. The first superconducting materials to be discovered displayed the effect near **absolute zero**. In recent years materials that become superconducting at ever higher temperatures have been developed, and the achievement of room-temperature superconductivity is within sight.

surface tension The attraction between **molecules** on the surface of a liquid. The effect is rather like a surface skin holding a liquid sample together. The cause is in fact **cohesion**; this tends to give particles near the surface a net inward force, as shown in the diagram.

The value of surface tension falls as temperature rises. This is partly why a warm liquid evaporates more quickly than a cool one. Surface

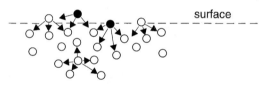

surface tension *There is a net inward force on particles near the surface of a liquid.*

tension also relates to **capillarity**; it explains why a small drop of liquid tends to take the shape of a sphere.

switch A device designed to control whether a **current** flows in a **circuit** or not. There are many designs of such on/off switches; the more complex ones let the user control the current in more than one circuit and/or change the track of a current. Others again (like dimmers and volume controls) can change the size of the current as well as turning it on and off. The basic action of a **transistor** is as an electronic switch.

symbol A character or a special shape that is agreed to have a particular meaning. Thus, in science, it is agreed to use I for current, kPa for kilopascal, K for potassium, and ♀ for female (also for the planet Venus). A symbol is not the same as an abbreviation, though sometimes symbols originated as abbreviations. Full stops should not be used with symbols. The precise use of capital and small letters is crucial – pA (picoamp) is not the same as Pa (pascal); mA (milliamp) differs a great deal from MA (megamp). *See also* **variable**, *Appendix A*.

T

table **Data**, mostly in numeric form in science, set out as a two-dimensional pattern of rows and columns. It is much easier to understand data in this way than when they appear as a set of sentences; it is also easier to see patterns in the data when in a table. Even better than a table may be a **graph**; however, it is common to put results in table form before plotting a graph. On the other hand, a table may be able to hold much more data than a graph can show; it also provides precise values where the user needs those.

telephone A system which carries speech and data as electric signals in a wire, as light signals in a fibre, or as radio signals. The standard handset, shown in the diagram, includes a carbon **microphone**, (a), and a simple **loudspeaker**, (b). The **sound** waves vibrate the microphone plate; this varies the resistance of the carbon conductor, so the constant applied voltage gives a varying current.

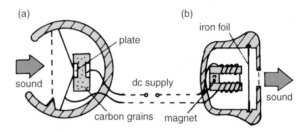

telephone *The mouthpiece (a) and earpiece (b) of a standard handset.*

The earpiece (speaker) combines a magnet with an electromagnet. The changing current in the electromagnet creates a varying magnetic field that makes the iron foil vibrate to produce sound.

telescope An **optical instrument** giving enlarged images of distant objects. The *Galilean telescope* (as used in opera glasses) combines a diverging **lens** and a converging lens. The standard *refractor* has two converging lenses; more magnification results from this arrangement, but the tube is longer and the image is upside down. The *reflector* has a converging **mirror** instead of the object lens (the lens nearer the object). The standard *radio telescope* has a design similar to the optical reflector's, but detects radio waves. *See also* **prism binoculars**.

television System of transmitting images by **radio** waves. Sound and computer data are carried in the same way. Even if not in colour, the transmission of visual data requires high-frequency waves because of the high information content. Video electronic circuits are therefore complex.

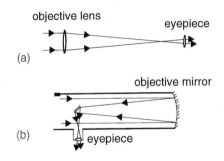

telescope *(a) Refracting astronomical telescope; (b) reflecting (Newtonian) telescope.*

Video signals can be carried by cables and glass fibres as well as by radio waves.

A television screen is one end of a **cathode-ray tube**. In a black-and-white set a fast-moving spot of light, created by a beam of electrons, scans across the screen, its brightness varying as it does so. In a colour set three electron beams, directed onto dots of three kinds of phosphorescent material, create red, green and blue spots of light. They build up a picture on the screen as a series of lines. In the system used in the UK, 25 different images are displayed each second. This would create noticeable flicker, except that each is displayed in two stages – first the odd-numbered lines, then the even-numbered ones, so that the whole screen surface is scanned 50 times per second.

temperature (*T*) The 'degree of hotness' or 'energy level' of a substance. It relates to the mean energy of the particles of the sample in question. Many types of **thermometer** exist to measure it, though none can measure mean particle energy as such.

A number of temperature scales are used. In daily life the *Celsius (centigrade) scale* is used, though in Britain and the USA the *Fahrenheit scale* is still quite common. In physics the **Kelvin scale** *(absolute or thermodynamic scale)* is of great value. The table shows the values of the two main **fixed temperatures** on these scales.

See also **absolute zero, kinetic model**.

Scale	Unit	Ice	Steam
Celsius	degree C, °C	0°C	100°C
Fahrenheit	degree F, °F	32°C	212°F
Kelvin	kelvin, K	273K	373K

tension The force in a stretched object. It is the effect of **cohesion** between the particles in solid and liquid materials: it tends to prevent their separating when pulled apart. *See also* **elasticity**.

terminal speed or terminal velocity The final constant speed of a body falling through a resistive fluid. The friction acting on the body increases with speed and at the terminal speed the downward force due to gravity equals in size the upward resistive force (**drag**). The net force on the body is zero, i.e. the body is acted upon by two **balanced forces**. This means that according to the second of **Newton's laws of motion**, the body is not accelerated, so carries on at a steady speed.

theory A concept, or set of concepts, to explain some aspect of the observed world. A theory (or **hypothesis**) may come to be called a **model**, a description, or a **law** if it survives long enough.

thermal capacity or heat capacity (C) The ability of an object to absorb thermal energy: the energy involved in changing its temperature by one degree. The unit is the joule per degree, J/°C or J/K.

The **specific thermal capacity** (c) of a substance is the energy needed to raise the temperature of one kilogram of a substance by one degree. The unit is the joule per kilogram per degree, J/kg°C or joule per kilogram per kelvin, J/kgK. *See also* **latent heat**.

thermal radiation The spectrum of electromagnetic radiation emitted by a hot object. This follows a definite pattern with a peak value at a frequency determined by the temperature of the body. It is also called black-body radiation, since a **black body** is not only a perfect absorber of radiation but also a perfect emitter.

thermionic effect The emission of **electrons** from a heated surface. This effect is used to provide the electron beam in cathode-ray tubes (in TV sets, computer monitors and cathode-ray oscilloscopes). In each case, the electrons are accelerated towards a phosphorescent screen by a high positive voltage on an **anode**. *See also* **emission**.

thermistor A device whose **resistance** falls as temperature rises, and therefore falls with increase of **current**. The device, a rod of certain metal oxides, is used in electronic circuits and as a thermometer.

thermocouple *See* **thermometer**.

thermodynamics The study of the interaction between **work**, **heating** and other energy transfer processes which change the **internal energy** of an object. Thermodynamics developed as a result of the invention and use of **heat engines**, but is of great importance in many applications of science where energy is involved, from the working of a living cell to the behaviour of the universe (**cosmology**).

thermometer A device used to measure **temperature** or 'hotness'. This relates to the mean energy of the sample's particles; a direct measure of this cannot be made. Therefore all thermometers measure something which depends on temperature, e.g. the length of a column of liquid in a tube, as in diagram (a), the pressure of a gas, the resistance of a length of wire or the voltage between the two junctions of a pair of metals (a *thermocouple*), (b). In each case the device is calibrated at certain **fixed temperatures**; it is then marked in degrees according to the scale used.

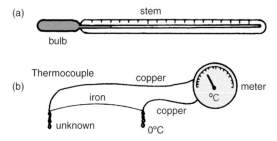

thermometer *Temperatures measured by using (a) the expansion of a liquid in a tube; (b) the voltage between two metals.*

thermostat A device used to keep the temperature of an object or room within a set range. Thermostats appear in many types of equipment in the home – cookers, irons and freezers, for instance. Many thermostats use a **bimetal strip**.

tides The cyclic rise and fall of sea level as a result of the gravitational pull of the Moon and the Sun. There are usually two high tides and two low tides in each lunar day; tides are higher at full moon and new moon, when Earth, Moon and Sun are aligned.

time The ordering and duration of events. Most living things have internal 'biological clocks', usually geared to **day and night** and the **seasons**. The unit of physical time, the second, was previously defined in terms of the day (the rotation of the Earth) or the year (the orbital movement of the Earth around the Sun). It is now defined in terms of the **frequency** of specified electromagnetic **radiations** given out by the common type of caesium atom.

torque (T) *See* **moment**.

torsion force A force that tends to twist a body. *See* **moment**.

total internal reflection The total **reflection** of radiation back into a medium when it meets a medium with a smaller **refractive constant** within a certain range of angles. Light passing from medium 2 (e.g. glass) to medium 1 (e.g. air) bends away from the normal (that is, the perpendicular), as shown in the diagram at (a). Thus there must be a stage, (b), at which a ray leaves at 90°. The angle of incidence, *i*, in this case is called the *critical angle, C*. The refractive constant, *n*, for medium 2 relative to medium 1 is given by:

$$n = 1/\sin c$$

If *i* is greater than *C* all the light must be reflected, as shown in the diagram at (c).
   See also **periscope, prism binoculars**.

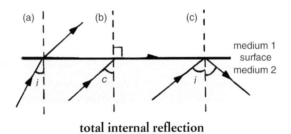

total internal reflection

tracer A small quantity of a radioactive **isotope** used to give scientists valuable information about a mechanical or biological system; the path of a tracer through a system can be checked using a detector because of the radiation it gives out. River pollution, sand movement in estuaries, blood flow and bearing wear can be investigated by using small quantities of radioactive tracers.

transducer A device that changes energy from one form to another. Microphones, **photocells** and loudspeakers are examples of transducers.

transformer A device whose normal use is to transfer electrical energy with a change in voltage between input and output. (The full name is *voltage transformer*.)
   Input **alternating current** causes an alternating magnetic field in the core. (This is built of thin strips to reduce **eddy currents**.) The changing field induces a changing voltage between the ends of the secondary (output) coil. *See* **electromagnetic induction**.

The ratio of the output voltage to the input equals the ratio of the numbers of turns (N) in the output and input coils: $V_2/V_1 = N_2/N_1$.

If there is 100% **power** transfer, $P_2 = P_1$, or $V_2 I_2 = V_1 I_1$. Thus, for a perfectly efficient transformer the ratio of the input and output currents equals the ratio of the output and input turns: $I_1/I_2 = N_2/N_1$.

In practice no transformer is 100% efficient, though values very close to this are common.

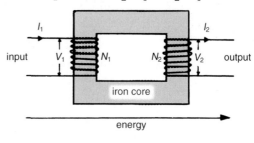

transformer

The main waste of energy comes from eddy currents in the core, the resistance of the coils, and the energy needed to switch the core **domains** during each cycle of magnetization.

For a 'step-down' machine, $V_2 < V_1$; then $N_2 < N_1$, and $I_2 > I_1$. Thus the output wiring needs to be thicker than that of the input. In the case of a 'step-up' machine, where $V_2 > V_1$, the reverse applies.

transistor A **semiconductor** device consisting of a combination of n-type and p-type semiconductors. It can be used to amplify signals. In the case of the bipolar or junction transistor (*see diagram*), a small signal applied to the base causes large variations in the current flowing between emitter and collector. In the case of the field-effect transistor (FET), the signal is applied to the gate and varies the current flowing from source to drain. *See* **amplifier**. The transistor can also be used as a **switch**, or in an **oscillator** circuit.

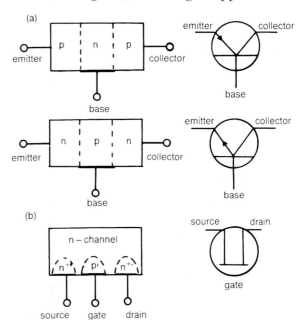

transistor (*a*) *Bipolar transistors;* (*b*) *field-effect transistor.*

transparent Able to let radiation through without significant **absorption**. No substance is fully transparent; even very clear glass absorbs some light.

truth table A table of the outputs of a logic gate for all its possible inputs. *See* **gate**.

turbine A device for extracting energy from a moving fluid. It consists of a set of curved blades on a shaft which are made to rotate by the force of the moving fluid. Power stations use turbines turned by the force of expanding steam to generate electricity. Similarly, water is used in hydroelectric power stations and wind is used in the advanced 'windmills' in *wind farms*.

U

ultrasonics Science and technology of *ultrasound*, **sound** waves of higher frequency than the human ear can detect (above 20 kHz). Some other animals can hear some ultrasonic ranges. The **sonar** system of bats, for instance, typically works at 50 kHz or higher. Ultrasound of megahertz frequencies has many uses in research, industry and medicine, e.g. for cleaning objects without damage, breaking up kidney stones, detecting mines at sea, and imaging babies in the womb. *Compare* **infrasound**.

ultrasound *See* **ultrasonics**.

ultraviolet A region in the spectrum of **electromagnetic waves**. The wavelength range is approximately 10^{-10} to 7×10^{-7} m, with a corresponding frequency range of 3×10^{18} to 4×10^{14} Hz. Produced by white-hot objects and certain gas discharges, ultraviolet radiation (UV) affects photographic film, and causes **fluorescence** and **photoelectric effects**. Human skin makes vitamin D using energy supplied by this radiation, and UV makes light skins tan. Especially at the higher frequencies, ultraviolet presents a hazard to eyes and skin. UV is also used in sterilizing equipment in hospitals, to kill germs. *See also* **ozone layer**.

units Standard quantities with respect to which all other quantities are measured and quoted. In the past, people devised units as they needed them. However, a true *system* of units must be: simple, causing no problems in use, and easy to learn and recall; coherent, with units linking neatly with each other; and widely accepted.

The first attempt to set up such a unit system was made in 1791; the *metric system* then devised was based on the centimetre, gram and second. It has been developed into *SI*, the Système International d'Unités; this is widely used in science today.

In SI, the base units include the metre, m (length), the kilogram, kg (mass) and the second, s (time). Units derived from these include those of density (the kilogram per cubic metre, kg m^{-3} or kg/m^3); momentum (the kilogram metre per second, kg ms^{-1} or kg m/s); and acceleration (the metre per second per second, ms^{-2} or m/s^2). Many derived units are rather complex. For example, the derived unit of pressure is the kilogram metre per second per second per square metre $-$ (kg m/s^2)/m^2. It is normal to give these special names, in this case the pascal, Pa.

Thus a set of *standard units* is obtained. If any of these is too large or too small for a given usage, a *prefix* is added to it. Each prefix describes how the new unit formed relates to the standard. Thus the kilometre is 1000 m, the kilovolt is 1000 V, and the kilonewton is 1000 N.

See Appendix A.

universal gravitational constant *See* **gravity**.

universe The system of all that exists. The universe started with the **Big Bang**, an explosion in which space, time, matter and energy were created about fifteen thousand million years ago. Since then the universe has expanded and developed into its present structure of **galaxies** grouped into clusters and superclusters. Each galaxy is a vast system of stars rather like the **Sun**, together with gas and dust. *See* **Hubble constant**.

upthrust The upward force on an object fully or partly submerged in a fluid. Its cause is the **pressure** difference between higher and lower levels. This upward force makes underwater objects lighter than in air. It also causes ships and balloons to float. *See also* **Archimedes' principle**.

uranium The element of **proton** number 92, the highest found naturally on the Earth. Uranium is a metal, with three natural **isotopes**, as shown in the table. All three are **radioactive**, breaking down through **alpha decay**.

In addition, uranium-235 is *fissile* – its nuclei can undergo **fission** when they absorb neutrons. As each fission leads to several more neutrons, a **chain reaction** may take place. This is the basis of **nuclear power** and the **atom bomb**.

Isotope	Z	N	A	$T_{1/2}$/Year	Abundance
uranium-234	92	142	234	3×10^5	0.006%
uranium-235	92	143	235	7×10^8	0.7%
uranium-238	92	146	238	5×10^9	99.3%

In the table, Z is the proton number, N is the neutron number, A is the nucleon number, and $T_{1/2}$ is the **half-life** period.

Whether for use as a fuel or in bombs, it is common to *enrich* the pure uranium, i.e. to increase the fraction of uranium-235 by a factor of several times beyond the usual 0.7%.

V

vacuum Strictly, a space that contains no matter. Since this is not achievable, the term is more generally used to refer to a region in which there are very few particles, such as that inside a **cathode-ray tube**. A cubic metre of air at standard **pressure** and **temperature** contains some 10^{25} particles; the figure for a good laboratory vacuum is about 10^{14}. Between the Earth and the Moon the value is far lower, and between galaxies lower still. However, the vacuum of outer space contains energy in the form of electromagnetic **radiation** and gravitational and electromagnetic **fields**.

valve 1. A device that allows a flow of gas, liquid or electricity in one direction only.

2. An **electron tube**, an electronic device that is based on the conduction of electricity through a **vacuum** and allows current to pass only one way.

Van de Graaff machine A machine that can produce a very high voltage by electrostatic means.

In use, electrons are collected from a moving belt by a sharp-edged metal comb, by the *action of points* (*see* **charge distribution**). The belt is made of a very good insulator and carries the positive charge remaining to another comb connected to the inside of a metal sphere. Here the action of points 'sprays' electrons on to the belt, and as a result the sphere becomes more and more positively charged until the insulation of the air breaks down. This means that ions are produced.

Van de Graaff machine

Discharge in the form of a spark may then occur, or there may be a fairly steady leakage.

Maximum voltage values as high as 500 kV are common. The very large models used in research may give voltages up to 10 MV. Small machines do not need a large voltage supply as the initial charge is provided by friction.

The device is named after the American physicist Robert Van de Graaff (1901–1967) who invented this device and developed it further for use in accelerating particles and generating X-rays.

vapour Matter in the gas **state** which can be made liquid by pressure. Above the substance's **critical temperature**, T_c, this cannot be done; the substance is then a true gas. Any solid or liquid evaporates to some extent at any temperature. The *saturated vapour pressure* (svp) of a substance is the pressure exerted by the vapour in equilibrium with the liquid. How svp varies with temperature is shown in the diagram.

This graph is very useful, as the same curve occurs in a number of contexts. As shown, it relates the svp, p, of the sample to its temperature, T. It can also show:

- how the boiling temperature T depends on the outside pressure p (*see* **pressure cooker**);
- the pressure p needed to liquefy the vapour at temperature T;
- whether a substance is liquid, vapour or gas in a given pressure/temperature situation (*see* **state change**).

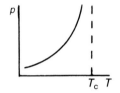

vapour *Graph showing how svp varies with temperature.*

variable Any item of **data** which is not a constant; in other words, a measure whose value may change. For example, π, the ratio of the circumference of a circle to its diameter, is constant in value (3.14159 . . .); however, circumferences and diameters are variables; they are not constant in value but vary with the circle concerned.

In an experiment it is usually important to find out how variables are related to each other. One quantity (the *independent variable*) is varied so that it can be seen what happens to another (the **dependent variable**). When it is shown how they relate in an equation (in a formula), a standard **symbol** is used for each one. The common symbols for general variables (ones with no special meaning) are x, y and z.

vector A measure which involves both size and direction. Examples are **velocity, acceleration, momentum, field strength**. *Compare* **scalar**.

velocity (v) The displacement of an object in unit time in a particular direction. The unit is the metre per second, m/s. To find velocity, the displacement s is divided by the time t: $v = s/t$. Velocity is a vector – *compare* **speed**. *See* **equations of motion**.

velocity ratio *See* **machine**.

velocity, relative The **velocity** of each of two objects moving relative to one another, that is, in each case, how one object seems to move when viewed from the other. *See also* **Doppler effect**.

vernier A device which allows a more precise reading of length than a ruler.

vibration A regular to-and-fro motion. Examples are the motions of a ruler 'twanged' over the edge of a table and of a pendulum bob. Such motions are common in physics. *See* **sound sources**.

In each case, the motion can be shown as a pure sine wave (*simple vibration* or *simple harmonic motion*) or as the sum of a number (perhaps very large) of sine waves.

viscosity or fluid friction A measure of a **fluid's** reluctance to flow. Syrup, for example, is more viscous than water. A good car engine oil is *viscostatic*, i.e. its viscosity does not change much as the temperature changes. An object falling through a viscous medium accelerates until the downward **gravity** force equals the upward friction force. It then falls at constant **terminal speed**.

vision The special sense for light that gives larger animals an image of the outside world. The sense organ used is the **eye**. In human beings, the normal eye can focus light from any object more than 250 mm away. *See also* **colour vision, vision defects**.

vision defects Faults in the eye which reduce its efficiency or ability to make a clear image. A *short-sighted* eye can focus on closer objects, but cannot see clearly for distances over a few metres. A diverging lens can correct the fault. A *long-sighted* eye cannot focus on close objects. A converging lens will correct the problem. *See also* **astigmatism**.

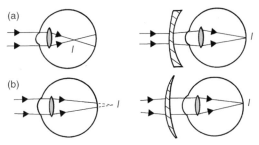

vision defects *Correction of (a) short-sightedness by a diverging lens and (b) longsightedness by a converging lens.*

volt *See* **voltage** *and Appendix A (SI units).*

voltage (V) A common name for either **potential difference** or **electromotive force**; the unit of both is the volt V. Electromotive force (emf) drives electric charges round a circuit. Potential difference is the voltage produced across an electrical device by, for example, current flowing through it.

One volt is the potential difference between two points when 1 joule of energy is absorbed or given out as 1 coulomb of charge moves between the points. Voltage is given by:

$$V = \frac{E}{Q}$$

volume (V) The space taken up by a quantity of matter. The unit is the cubic metre, m^3. The volume of a brick-shaped sample is the product of length, width and height. That of a sphere is $\frac{4}{3}\pi r^3$, where r is the radius. The volume of a gas sample depends on the temperature and pressure (*see* **gas laws**). On the other hand, the volumes of solid and liquid samples are fairly constant.

voltage divider *See* **potential divider**.

voltmeter A device used to measure **voltage** or **potential difference**. Moving-coil voltmeters are similar in design to **ammeters**; however, voltmeters measure the potential difference across a circuit component and so must be connected in parallel with it, and they have a high resistance so that they draw a negligible amount of current.

Digital voltmeters are now in general use, in which voltage is measured electronically by, say, a magnetic force, the result being shown on a numerical display, in contrast with the mechanical voltmeter in which the result is shown by the movement of a pointer across a scale.

W

water, anomalous expansion of The expansion of water as it freezes. Most liquids increase in density and contract as they cool, and indeed water does this down to about 4 °C, but as it cools further it expands and becomes less dense. Graphs (a) and (b) illustrate these effects. It expands even more when it turns to ice. This effect has important practical and environmental consequences: water freezing in pipes will tend to burst them as the ice and trapped water expand; the coldest water in a pond floats to the surface to form ice, which then acts as an insulator so that deeper water stays above freezing for long periods, allowing organisms to survive.

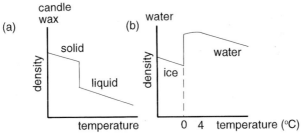

water, anomalous expansion of *Density/ temperature graph of (a) candle wax and (b) water.*

water cycle The transfer of water between land, seas and sky. Liquid water evaporates from the surface into the atmosphere, where it may form clouds. It precipitates from the atmosphere as rain or snow to become surface water again. Streams and rivers (some under the ground) carry the liquid to the seas.

At any moment, less than 1% of the Earth's water is in the air and in streams, lakes and rivers. About 2% is in the form of ice, while the rest is in seas and oceans.

watt (W) The unit of **power**, the rate of energy transfer.

wave A disturbance that travels through space or a medium as a series of oscillations. The production and transmission of a wave involve **vibration**. The basic *wave equation* $c = f\lambda$ relates the wavelength λ and the frequency f of a wave to its speed c in a medium.

Most waves are transverse or longitudinal. *Transverse waves* consist of vibrations whose direction is across the direction in which the wave travels.

With *longitudinal waves* the vibration direction is the same as the direction in which the wave travels. The diagram shows the two types in the case of a long spring.

All waves can show **absorption**, **reflection**, **refraction**, **diffraction** and **interference** effects. **Polarization** can occur only with transverse waves. *See also* **electromagnetic waves, sound**.

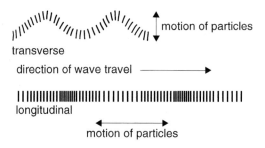

wave *Particle motion in transverse and longitudinal waves.*

wavelength (λ) The distance taken up by a single **cycle** of a wave. The unit is the metre, m.

The wavelength, λ, the frequency, f, and the speed, c, of a wave in a medium are related by the *wave equation*, $c = f\lambda$. If the wave passes into a second medium (**refraction**), the speed will change; as frequency is fixed, λ will also change.

weather The condition of the atmosphere near the Earth's surface, in particular as it changes from moment to moment and from place to place. The main factors are the air pressure and moisture content. These affect the flow of air (wind) and water precipitation (e.g. fog, frost, rain). *See also* **climate**.

weathering The breakdown of **rock** on the Earth's surface by mechanical processes (such as the expansion of ice in cracks), or by the dissolving of chemicals in water, or the action of living things such as plants, animals and people. The fragments are broken or worn into smaller and smaller pieces and eventually become sand and soil. *Compare* **erosion**.

weight (W) The gravitational force on an object. Weight is a **force**, so its unit is the newton, N. The value depends on the object's **mass**, m, and on the gravitational field strength, g (*see* **free fall**): $W = mg$. This gives, as a unit for g, the newton per kilogram, N/kg.

An object's mass is constant in normal circumstances. However, the value of g depends on where the object is. Thus, weight depends on position. On the Earth g = 9.8 m/s^2 or 9.8 N/kg. For most calculations, g is taken to be 10 m/s^2 or 10 N kg^{-1}. Thus the weight of a 10-kg object is approximately 100 N. On the moon (g = 1.5 m/s^2 or 1.5 N/kg) its weight would be approximately 15 N. *See also* **weightlessness**.

weightlessness The condition of an object in **free fall**, in which the usual perceptible effects of weight are absent. A spacecraft in orbit, with its engines turned off, is in free fall; it constantly accelerates at g. All objects in the cabin accelerate at the same rate – so they all move together and 'float' as though they had no weight, though there may actually be strong gravitational forces on them.

white dwarf *See* **star, Herzsprung–Russell diagram**.

wind A volume of air moving over the Earth's surface. Winds are the result of changes of air pressure, the effect in turn of different temperatures in different places. High ground and the spin of the Earth affect them once started.

wind farm A collection of wind-driven **turbines** for generating electricity from the force of moving air in a wind. Wind farms are an example of the use of **alternative energy**.

work (*W*) A very common type of **energy** transfer in physics. Work is done when a force (F) moves a distance (s) in the direction of the force:

$$W = Fs$$

The unit of energy, the joule (J), is defined in terms of work:

work done (energy transferred) by 1 N moving 1 m = 1Nm $\equiv$ 1J.

If there is no displacement in the direction of the force, there is no energy transfer/work done. An example is the force of **gravity** holding an object in circular orbit. The force acts towards the centre of the circle; s is zero in that direction as the object does not move closer to the centre.

X – Y – Z

X-rays A region in the spectrum of **electromagnetic waves**. The wavelength range is approximately 10^{-12} to 10^{-10} m with a corresponding frequency range of 3×10^{20} to 3×10^{18} Hz. X-rays are produced in *X-ray tubes* by the impact of high-speed electrons on metals.

Young's two-slit experiment An experiment first carried out by Thomas Young in 1801, which showed that light had the characteristics of a wave. A parallel beam of light is passed through two narrow slits placed close together. Each slit diffracts (spreads out) the light into two overlapping beams. Where they overlap the light waves interfere with each other to produce a pattern of light and dark bands called *Young's fringes*, which can be viewed directly or on a screen. *See also* **interference, diffraction**.

Zener diode breakdown diode A type of **diode** used to keep the applied voltage of a circuit within a set range. If the voltage becomes too large, a sudden electrical breakdown takes place in the diode which makes it conduct, so limiting the voltage rise.

zodiac A broad belt in the sky which contains the paths of the Sun, Moon and planets. It was divided some thousands of years ago into 12 zones, corresponding to the 12 months of the year, each characterized by the major constellation in the zone, such as Aries, Leo, Gemini etc. (the signs of the zodiac). However, the very slow wobble of the Earth's axis has changed the seasons so that the conventional zodiacal signs are now out of step with the constellations that the Sun is actually in at the corresponding times.

Appendix A

Physical Units

SI Units Physical quantities are measured in a standard system, the *Système International d'Unités* or *SI System*. There are seven base units; all others are combinations of these. The seven base units are:

the metre	m	the unit of length
the kilogram	kg	the unit of mass
the second	s	the unit of time
the ampere	A	the unit of electric current
the kelvin	K	the unit of (thermodynamic) temperature
the mole	mol	the unit of amount of substance
the candela	cd	the unit of luminous intensity

Many units are named after the scientists who worked in that area of physics. In this case the unit uses the name with a lower case initial letter but uses the initial letter in capitals for the abbreviation, e.g. ampere (A), kelvin (K), newton (N).

Derived units Most physical units are derived units, made up of combinations of the base units. This is because the units link together different quantities. For example, speed links length and time:

$$\text{speed} = \frac{\text{distance}}{\text{time}} = \frac{\text{length units}}{\text{time units}}$$

so the unit for speed (and velocity) is m/s. Similarly, acceleration is m/s^2. Sometimes derived units are shown using the following style:

$$m\ s^{-1}, m\ s^{-2}\ \text{etc}$$

The unit of force, the newton, links together mass and acceleration via the Newton formula $F = ma$. Thus the unit of force is:

$$(\text{mass unit}) \times (\text{acceleration unit})$$
$$\text{i.e. kg m/s}^2$$

Because of its importance this unit is given the special name of the *newton*.

The numerical values of all physical quantities should be given with the correct unit. Note that some quantities have no units because they represent the ratio of similar quantities, e.g. refractive index (ratio of speeds or sines), efficiency (ratio of energies).

Multiples and sub-multiples These are used when the standard unit is inconveniently large or small for the quantity being considered, so that using it would mean the use of very large or very small numbers. A prefix is placed in front of the unit name. Thus the wavelength of red light at 6×10^{-7} m is often written as 600 nanometre (nm). The distance between two towns is written as 250 km rather than 250,000 m.

Standard prefixes

multiples (larger)		by a factor	submultiples (smaller)		by a factor
deca-	da	10	deci-	d	10^{-10}
hecto-	h	100	centi-	c	10^{-20}
kilo-	k	1000	milli-	m	10^{-30}
mega-	M	1,000,000 (10^6)	micro-	μ	10^{-60}
giga-	G	10^9	nano-	n	10^{-90}
tera-	T	10^{12}	pico-	p	10^{-12}
peta-	P	10^{15}	femto-	f	10^{-15}
exa-	E	10^{18}	atto-	a	10^{-18}

The most common units in physics:

Space and time

quantity	unit	conventional symbol in formulae
breadth	m	b
height	m	h
depth	m	d
thickness	m	d
radius	m	r
distance along path (displacement)	m	s
wavelength	m	λ
area	m^2	A
volume	m^3	V
time	s	t
period	s	T
frequency	Hz (1/s)	f, ν

Motion

speed	m/s	u, v
velocity	m/s	u, v
acceleration	m/s^2	a
acceleration of free fall	m/s^2	g

Mechanics

mass	kg	m, M
density	kg/m^3	d, ρ
force	N	F
pressure	pascal (Pa)	P
tensile strength (Young modulus)	pascal (Pa)	E
tensile stress	pascal (Pa)	σ
coefficient of friction	*none*	μ
momentum	kg m/s; Ns	P
impulse	kg m/s; Ns	I
gravitational field strength	N/kg	g
weight	N	W

Electricity

current	A	I
charge	coulomb (C)	Q
potential difference	volt (V)	V
electromotive force	volt (V)	E
resistance	ohm (Ω)	R
resistivity	ohm-metre (Ω m)	ρ
capacitance	farads	C
electric field strength	V/m or N/C	E

Atomic and nuclear physics

atomic number (proton number)	*none*	Z
mass number (nucleon number)	*none*	A
relative atomic mass	*none*	A_r
activity of radioactive source	becquerel	s
half life	s	$T\frac{1}{2}$ or $t\frac{1}{2}$

Molecular quantities

relative molecular mass	*none*	M_r
amount of substance	mole (mol)	n
molar mass	kg/mol	M
Avogadro constant	mole^{-1} (mol^{-1})	L

Thermodynamics

temperature (absolute, thermodynamic)	kelvin (K)	T
Celsius temperature	°C	t or θ
energy	J	E
work	J	W
power	W (J/s)	P
heat (internal energy)	J	Q, q
heat capacity	J/K	C
specific heat capacity	J/kg K	c
specific latent heat	J/kg	L
thermal conductivity	W/m K	λ

Other

refractive index	*none*	n

Appendix B

Selected physical constants

In general, the constants are quoted to two significant figures.

Fundamental constants

constant	symbol	value
speed of light	c	3.0×10^8 m/s
charge on electron/proton	e	$\pm 1.6 \times 10^{-19}$ C
gravitational constant	G	6.7×10^{-11} N m²/kg²
Faraday constant	F	9.6×10^4 C/mol
electron rest mass	m_e	9.1×10^{-31} kg
proton rest mass	m_p	1.7×10^{-27} kg
Planck constant	h	6.6×10^{-34} Js
gas constant	R	8.3 J/K mol

Other useful constants and quantities

standard atmospheric pressure	P	1.0×10^5 Pa (1 bar)
astronomical unit (distance Earth-Sun)	AU	1.5×10^{11} m
light-year	ly	9.5×10^{15} m (6.3×10^4 AU)
parsec	pc	3.1×10^{16} m (3.3 ly)
radius of Earth		6.4×10^6 m
mass of Earth		6.0×10^{24} kg
standard Earth gravity		9.8 N/kg or m/s²
mass of Moon		7.4×10^{22} kg
mass of Sun		2.0×10^{30} kg
luminosity of Sun		3.9×10^{26} W
Hubble constant		about 15 km/s/Mly

$1 \text{ m}^2 = 10^4 \text{ cm}^2$
$1 \text{ m}^3 = 10^6 \text{ cm}^3$
$1 \text{ litre} = 1000 \text{ cm}^3 = 10^{-3} \text{ m}^3$
$1 \text{ g/cm}^3 = 1000 \text{ kg/m}^3$
$1 \text{ year} = 31.56 \text{ Ms}$
$1 \text{ kWh} = 3.6 \text{ MJ}$
$1 \text{ eV} = 1.6 \times 10^{-19} \text{ J}$

density of water	1000 kg/m^3
specific heat capacity of water	4200 J/kg K
specific latent heat of water/steam	$2.3 \times 10^6 \text{ J/kg}$
specific latent heat of water/ice	$3.3 \times 10^5 \text{ J/kg}$
specific heat capacity of copper	390 J/kg K
specific heat capacity of lead	128 J/kg K
specific heat capacity of glass	600 J/kg K

Appendix C

Formulae and relationships

Kinematics – the study of motion

acceleration $\qquad a = \frac{v-u}{t}$

speed $\qquad v = \frac{s}{t}$

average speed $\qquad v_{av} = \frac{u+v}{2}$

distance $\quad$ s = average speed × time = $(\frac{u+v}{2})t$

distance travelled by a uniformly
  accelerated object $\qquad s = ut + \frac{1}{2}at^2$

  if time is unknown use: $\qquad v^2 - u^2 = 2as$

Dynamics – the study of forces and motion

force, mass and acceleration $\quad F = ma$

momentum $\qquad P = mv$

impulse $\qquad Ft = \Delta P = m\Delta v$

Gravitation

acceleration of free fall $\qquad g \ [\text{m/s}^2]$

gravitational field strength $\qquad g \ [\text{N/kg}]$

gravitational potential energy $\quad mgh$

gravitational force $\qquad F = G\frac{Mm}{r^2}$

period of a simple pendulum $\quad T = 2\pi\sqrt{(l/g)}$

Matter

density $\qquad d = \frac{m}{V}$

weight $\qquad W = mg$

pressure $\qquad P = \frac{F}{A}$

pressure in a fluid $\qquad P = hdg$

gas law $\qquad \frac{PV}{T} = constant$

Boyle's law	$PV = constant$
elasticity	$F = kx$
strain	$\dfrac{\text{change in length}}{\text{original length}}$
stress	$\dfrac{F}{A}$

moment (torque)		$T = Fs$
area	square	$A = l^2$
	rectangle	$A = bl$
	circle	$A = \pi r^2$
	triangle	$A = \frac{1}{2}bh$
volume	cube	$V = l^3$
	cuboid	$V = lbh$
	sphere	$V = \frac{4}{3}\pi r^3$
	cylinder	$V = \pi r^2 h$

Energy and power

work	$W = Fs$ or Fd
kinetic energy	$E_k = \frac{1}{2}mv^2$
gravitational potential energy	$E_p = mgh$
energy transferred electrically	VIt or QV or $\dfrac{V^2 t}{A}$

change in internal energy $\Delta H = mc\Delta T$ or $\Delta E = mc\Delta\theta$
    [T is used for kelvin, θ for Celsius temperatures]

energy to change state $\Delta H = mL$

power $P = \dfrac{\text{work done}}{\text{time}}$ or $\dfrac{\text{energy transferred}}{\text{time}}$

electrical power $P = VI$ or I^2R or $\dfrac{V^2}{R}$

radiation energy, $H = hf$
    per photon

efficiency $\dfrac{\text{energy output}}{\text{energy input}} \times 100\%$

force ratio of machine $\dfrac{\text{load}}{\text{effort}}$
(mechanical advantage)

distance ratio (velocity ratio)	$\dfrac{\text{distance moved by effort}}{\text{distance moved by load}}$ or $\dfrac{\text{speed of effort}}{\text{speed of load}}$

Electric circuits

current	$I = Q/t$
resistance	$R = \dfrac{V}{I}$
voltage	$V = \dfrac{\text{energy}}{\text{charge}}$
circuit with internal resistance	$E = Ir + IR$
resistivity	$\rho = \dfrac{RA}{l}$
resistors in series	$R = R_1 + R_2 + R_3 + \ldots\ldots$
resistors in parallel	$\dfrac{1}{R} = \dfrac{1}{R_1} + \dfrac{1}{R_2} + \dfrac{1}{R_3} + \ldots\ldots$
capacitance	$C = \dfrac{Q}{V}$
capacitance of a plane capacitor	$C = \dfrac{kA}{d}$
capacitors in series	$\dfrac{1}{C} = \dfrac{1}{C_1} + \dfrac{1}{C_2} + \dfrac{1}{C_3} + \ldots\ldots$
capacitors in parallel	$C = C_1 + C_2 + C_3 + \ldots\ldots$
electric field strength	$E = \dfrac{F}{Q}$
electric field strength in uniform field	$E = \dfrac{V}{d}$

Electromagnetism

Faraday law	$V = $ rate of change of field
transformer rule	$\dfrac{V_1}{V_2} = \dfrac{N_1}{N_2}$

Waves

wave speed	$v = f\lambda$
frequency	$f = $ cycles or waves per second

period	$T = \dfrac{1}{f}$
two-slit interference; fringe separation	$T = \dfrac{\lambda D}{d}$

Optics

lens and mirror formula	$\dfrac{1}{f} = \dfrac{1}{u} + \dfrac{1}{v}$
curved mirror	$f = \frac{1}{2} r$
'power' of a lens or mirror	$P = \dfrac{1}{f}$
refractive index, air to a medium	$n = \dfrac{\text{speed in air}}{\text{speed in medium}} = \dfrac{\sin i}{\sin r}$
critical angle	$\sin C = \dfrac{1}{n}$

Note Rather than use many different symbols for related measures, subscripts are often added. These are numbers or letters below the line; their function is to show to what the measure refers. Thus v_1 and v_2 are two different speeds, while n_{1-2} is the refractive constant for waves passing from medium 1 into medium 2. R_1, R_2, and R_3 would be the resistances of three resistors. The resistors are sometimes called A, B and C; their resistances would then be R_A, R_B and R_C.

Appendix D
Circuit symbols

Conductors

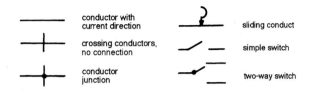

———	conductor with current direction	⌇	sliding conduct
—+—	crossing conductors, no connection	⁄ —	simple switch
—†—	conductor junction	⁄ ═	two-way switch

Sources

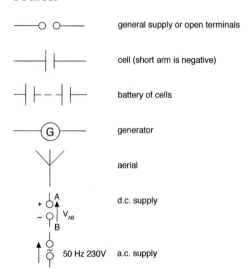

—○ ○—   general supply or open terminals

—| |—   cell (short arm is negative)

—| |– –| |—   battery of cells

—(G)—   generator

    aerial

+ ○ A, – ○ V_{AB} B   d.c. supply

○ ~   50 Hz 230V   a.c. supply

Passive devices

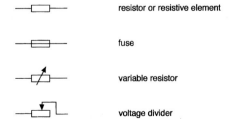

—▭—   resistor or resistive element

—▱—   fuse

—⌿—   variable resistor

—⌐_⌐—   voltage divider

Passive devices (continued)

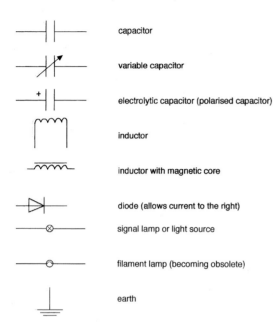

	capacitor
	variable capacitor
	electrolytic capacitor (polarised capacitor)
	inductor
	inductor with magnetic core
	diode (allows current to the right)
	signal lamp or light source
	filament lamp (becoming obsolete)
	earth

Active devices

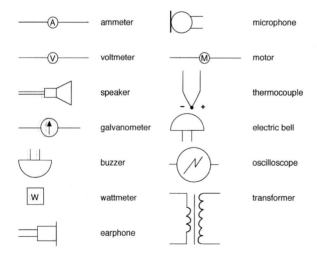

ammeter		microphone	
voltmeter		motor	
speaker		thermocouple	
galvanometer		electric bell	
buzzer		oscilloscope	
wattmeter		transformer	
earphone			

Electron (vacuum) tubes

  diode

  cathode ray tube

Semiconductor devices

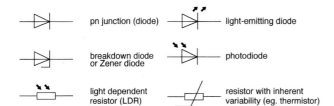

pn junction (diode)

light-emitting diode

breakdown diode
or Zener diode

photodiode

light dependent
resistor (LDR)

resistor with inherent
variability (eg. thermistor)

Sub-circuits

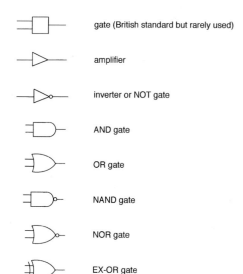

gate (British standard but rarely used)

amplifier

inverter or NOT gate

AND gate

OR gate

NAND gate

NOR gate

EX-OR gate

Appendix E

Data about the Planets

	Mercury	Venus	Earth
mean distance from Sun			
(10^6 km)	58	108	150
(Earth = 1)	0.4	0.7	1.0
Mass			
($\times 10^{23}$ kg)	3.3	48.7	59.7
(Earth = 1)	0.056	0.0815	1.0
Density			
(kg/m^3)	5400	5300	5500
Radius			
(km)	2439	6052	6378
(Earth = 1)	0.38	0.95	1.0
Period			
(Earth years)	0.24	0.61	1.0
Rotation period			
(hours)	1416	5832	24
Inclination of equator to			
orbital plane (degrees)	2	177	23
Mean surface temperature			
(degrees Celsius)	350	460	20

Mars	Jupiter	Saturn	Uranus	Neptune	Pluto
224	778	1427	2870	4497	5900
1.5	5.2	9.5	19.1	30.0	39.3
6.42	19,000	5680	866	1030	0.11
00.11	31.79	95.1	14.56	17.24	0.002
4000	1300	700	1270	1640	2000
3397	71,492	60,268	25,559	25269	1140
0.53	11.2	9.45	4.01	3.96	0.18
1.88	12	29	84	165	248
25	10	10	111	16	154
25	3	27	98	30	122
–55	–110	–180	–210	–220	–230

Appendix F

Geological time charts

eon	era	period
	Cenozoic	Quaternary
		Tertiary
Phanerozoic	Mesozoic	Cretaceous
		Jurassic
		Triassic
	Palaeozoic	Permian
		Carboniferous
		Devonian
		Silurian
		Ordovician
		Cambrian
Proterozoic		
Archaean		

epoch	began (millions of years ago)	life
Holocene	0.01	
Pleistocene	1.64	appearance of humans
Pliocene	5.2	
Miocene	23.5	
Oligocene	35.5	
Eocene	56.6	
Palaeocene	65	mammals flourishing
	146	dominance of dinosaurs
	208	first birds
	245	first mammals, dinosaurs
	290	
	363	first reptiles
	409	first amphibians
	439	first land plants
	510	first fish
	570	first fossils
	2500	earliest life
	4600	